BARROW STEELWORKS

Bessemer steelmaking. (Geoff Berry)

BARROW STEELWORKS

AN ILLUSTRATED HISTORY OF THE HAEMATITE STEEL COMPANY

STAN HENDERSON & KEN ROYALL

Cover illustrations: *Front cover main image*: courtesy of British Steel; *inset images and back cover*: Ken Royall collection.

First published 2015

The History Press
The Mill, Brimscombe Port
Stroud, Gloucestershire, GL5 2QG
www.thehistorypress.co.uk

British Library Cataloguing in Publication Data.
A catalogue record for this book is available from the British Library.

ISBN 978 0 7509 6378 7

Typesetting and origination by The History Press
Printed in Malta by Melita Press

Contents

Authors' Note

Limitations of space, and thereby cost, have mitigated against the inclusion of all information collated about the works and processes of the Barrow Haematite Steel Company (BHS Co.). More so regarding the Open Hearth process (chapter 6). It is hoped to correct this by the publication of another book – a companion to this current work – at some point in the future.

About the Authors

Ken Royall joined Barrow steelworks in April 1947 after being demobbed from the RAF, where he had been ex-WOp/AG radio operator, and started in the newly established Instrument Department as an instrument technician. Initially he found working in the steelworks like living in another world – most areas were extremely hot, dirty and potentially very dangerous – but it was a happy place in which to work. He became works photographer in 1957. Ken witnessed the dramatic changeover from orthodox steelmaking to the new continuous casting process. Following the demolition of the old plant he said that the working atmosphere became less frenetic but a lot cleaner. Ken eventually became an instrument and fuel engineer. His position, prior to redundancy in 1980, was department manager. Since retiring his main hobby, photography, has occupied most of his spare time.

Born and brought up in Hindpool, Stan Henderson comes from a family of steelworkers. His grandfather went into the rolling mills in 1911, and his parents, paternal uncles, friends and neighbours all worked at the steelworks. He developed a schoolboy fascination for the industry that could be likened to an addiction to steam trains. He worked at Barrow works for just twelve months, 1964–65, and said that walking through the gates for the first time was like *déjà vu*. He served an apprenticeship at Barrow shipyard and worked for twenty-five years in the Ship Drawing Office, moving latterly into quality assurance, before taking voluntary redundancy in 1994. He is now semi-retired and lives on Walney Island. He has been collecting information for the book since the works closed in 1983.

A furnaceman 'taking five'.

Taking the bath temperature of No.1 open-hearth furnace.

Acknowledgements

This book relies on references to and quotations from several publications. A bibliography of works consulted is included to allow the reader the opportunity for further reading.

I wish to acknowledge the invaluable assistance given by so many who directly or indirectly contributed. I am indebted to my late friend and colleague, George Hurley of the Ship Drawing Office of Vickers, Barrow. George provided my copy of the *Iron and Coal Trades Review*, which has been the main reference source for the early years of the works. He also related many amusing anecdotes – which could easily become the subject of another book.

In particular, I would like to thank Julie Bromley of Corus Rail, Workington who facilitated a visit to the Workington No.1 rail mill in 2002. Also Matthew Roylance for his editing skills, and all that such necessitated.

Grateful thanks to ex-Barrow steelworkers Derek Barnes, Albert Brennan, Jim Ducie, Tony Frankowski, Jim McGlennon, Peter Keenan, Ken Law, Ray Millard, Billy Miller, Bill Pearson, and Jim Walton, who between them have over 300 years' experience on which to draw.

Acknowledgements are made to the staff of Cumbria Archives and Local Studies Centre (Barrow), Bill Myers and the *Barrow Evening Mail* and Sabine Skae of the Dock Museum. The Barrow coat of arms is used under open data from Barrow Borough Council, 19 May 2011.

All photographs used in this book, unless stated otherwise, were taken by K.E. Royall, manager of the Fuel and Instrument Department (also the work's photographer from 1957), or sourced from his extensive collection. Every attempt has been made to contact copyright holders. Plate negatives (12x10) now reside at the Dock Museum.

Introduction

This book is an attempt to write a local industrial history, at grass-roots level, which may be of value to a new generation of students. It is not about the founding fathers or the rich mine owners of whom much has already been adequately documented. Neither is it an economic history. The book is about the buildings, plant and machinery that collectively comprised the once-mighty Haematite Steel Company at Hindpool. It is also about the men and women with brains and ingenuity and the men whose sweat and aching muscles ran the mills and furnaces of this once colossal works. The project has been an undertaking pursued by the authors in an attempt to preserve the memory of the once prominent local iron and steelworks, and the lives of some of the many who contributed to the works having the unique distinction of twice being a world leader in steelmaking technology nearly 100 years apart. The *Iron and Coal Trades Review* of 4 August 1899 noted:

> Barrow works of the first magnitude were erected to stand or fall by the manufacture of Bessemer steel, and primarily of the rolling of that steel into rails, which had been done nowhere else. It was a bold experiment. The conditions under which the rail manufacture was carried on were hardly sufficiently fixed to determine whether it would be entirely successful. The only rails previously made in England had been treated as cast steel and hammered from ingots 7 or 8 inches square and 4 feet long in four heats. It was some time later before the size of the ingot was raised to a sufficient size to make two thirty feet rails. By the 1870s ingots at Barrow are produced of 5 and 2-tons in weight, and the mills are capable of rolling rails 250 feet in length. The only limiting factor being the amount of metal the Hindpool furnaces could deliver.

Many years later, during the 1950s, the aforementioned ingots would be tossed onto the proverbial scrapheap when Barrow works developed a high-speed process for the continuous casting of steel billets and slabs. This high-speed process consigned several steps of orthodox steelmaking to the history books. Barrow once again played host to the world's steelmaking fraternity as delegates from steelworks as far away as Japan came to the town, in much the same way as they had in 1874 when Andrew Carnegie and the Iron and Steel Institute came to Barrow. The unfortunate thing was the achievement did nothing to boost the town's economy, or for jobs locally. Taking the broader view, the successful development work in Barrow was part of the biggest improvement in steelmaking since the Bessemer process itself. It may be of interest to note that the casting speeds achieved at Barrow during the 1950s are still an unbroken record.

The idea for this book was hatched in 1984, after having salvaged various items from a skip on the demolition site, which would be visited almost daily. I suppose that like many others I found it hard to accept that it would soon be gone, having grown up in the shadows of its tall chimneys, rumblings, screaming furnaces and general daily commotion. It was assumed that, like Black Combe in the background, it would stand forever.

There are still several gaps in our knowledge with regard to the early history of the works and then the interwar years. Unlike Workington Iron and Steel Company, very little has been recorded – or survived for posterity. By far the best reference source is the *Iron and Coal Trades Review* of 4 August 1899, the narrative of which was copied *verbatim* in the report of the Iron and Steel Institute visit to the town in 1903, then again in the 1937 book Barrow Steel. The reasons for the production of the latter publication have always been a mystery. This book, which makes reference to being a modern steelworks (which it certainly wasn't), was issued just two years before it was poised to close down – but for the onset of the Second World War it most certainly would have. Was this book, then, intended to stand as something of a memorial to a glorious past?

It is not possible in this type of book to give an account of all plant and machinery, which for a period of over 100 years was being constantly upgraded, replaced or rearranged; such a book would take on encyclopaedic proportions, or become just one large engineering inventory. The authors therefore have focused on items, hopefully, of historical significance.

During the thirty years spent collecting information it has not been possible to ascertain the location of the metal mixer, details of the slitting mill, tin bar mill or tyre mill. We do not know precisely when steelmaking by the Bessemer process ceased. It is a fact that whilst the place was an integrated steelworks it was not designed that way. The steelmaking side, east of the Furness Railway main line, was in fact a bolt-on, being laid down six years after the ironworks was established. This meant that the transporting of molten iron between blast furnaces and converters was not ergonomically sound, a journey of over 1 mile being involved for the ladle cars. At Moss Bay, Workington the same journey for the ladles was about 150yds.

S. Henderson, 2014

1

Hindpool

Queen Victoria died on 22 January 1901 and she had reigned as Queen of the United Kingdom since 1837 when she succeeded William IV. During her reign massive advances were made in technology and there was a worldwide expansion of boundaries due to the development of railways, heavy engineering and telegraphic communication. Her long reign spanned the second part of the Industrial Revolution and locally the period of what was Barrow's greatest development. The most significant early effects of this progress were in the areas of the Strand and Hindpool. Hindpool became home to many local industries, and by far the largest and most significant was the iron and steelworks, aided in no small measure by the Furness Railway Company. From the excellent 1958 work of Dr J.D. Marshall we learn that the land at Hindpool, which became the site of the iron and steelworks, was bought by the Furness Railway Company from the Cranke family of Urswick.

The site was leased to Messrs Schneider and Hannay for the erection of blast furnaces. The first two were laid down in 1857 and put into service in 1859. By 1860 there were four, seven by 1862 and ten by 1866. With the forming of the Barrow Haematite Steel Company in 1864, Schneider and Hannay's partnership was dissolved, with their assets at Hindpool, together with the mining interests in Furness, incorporated into the new company, of which they became directors.

By 1870 there were fourteen blast furnaces and eighteen Bessemer converters deployed in iron and steel production. The steelworks, which occupied the site to the east of the Furness Railway's main line, had a frontage on Walney Road of almost 1 mile and covered a vast area.

To cater for the rapid expansion of the town and the influx of workers, Hindpool became like one large building site. A London firm was contracted to build an estate of terraced houses. The steel company engaged a Scottish firm, Smith and Caird, to build a block of flats using local sandstone on a triangular piece of land adjacent to the works. These flats, because of their resemblance to Glasgow tenements, became known as the

Scotch Buildings. They housed about 950 people – more if they were well acquainted. Although an eyesore latterly, upon completion they must have looked quite impressive with their wide pavements of Coniston green slate around the external perimeter. These pavements were to survive for a few years after the original buildings disappeared. Demolition work began in May 1956 by John Binnell of Cameron Street, Barrow Island although the last vestiges managed to survive into early 1960.

The area known as Lower Hindpool had the largest concentration of ironworkers. These were the streets between Duke Street and Hindpool Road and in the early days of the works were occupied mainly by migrant workers. As iron and steelmaking was thirsty work this area boasted the most pubs and alehouses. The main watering hole of the ironworkers was the Hindpool Hotel, found on the corner of Hindpool Road and Blake Street and said to have the longest bar in the town. Steelworkers tended to favour the Queens Hotel, another large establishment with twelve bedrooms and stables to the rear, on the corner of Blake Street and Duke Street. Another popular venue was the Hammer and Pincers (later the New Inn) on the corner of Franklin and Steel Streets. Built by William Gradwell, this pub used to supply 'near beer' to the mill workers who, in those early days, could be working up to sixty hours per week. In later years the Wheatsheaf, on Hindpool Road, assumed the mantle of the area's best pub and was regularly frequented by steelworkers. Never before had there been such a coming together of wit, talent and characters, the pub could have easily been the last bastion of the Anacreontic Society in its celebration of wine, women and song!

Hindpool people referred to the flats as simply 'the buildings'. In 1933 the steelworks sold them to a housing association for £18,000.

View of the Scotch Buildings from the south *c.* 1955. Looking up Blake Street, St James's church and school can be seen in the distance; most of the construction costs, as well as the land, were paid for by the steelworks' directors. Just visible to the left, along Walney Road, is the entrance to Tay Street.

The Duke Street elevation. The Steelworks Club, which was on the corner of Duke Street and Walney Road, is to the right of the photo. (James Melville, courtesy of Alice Leach)

Demolition of the Scotch Buildings on 22 May 1956. Binnell's lorry is at the junction of Duke Street and Walney Road tugging at a section of roofing.

Two views of the Walney Road facade under the British Steel Corporation banner, 1970s.

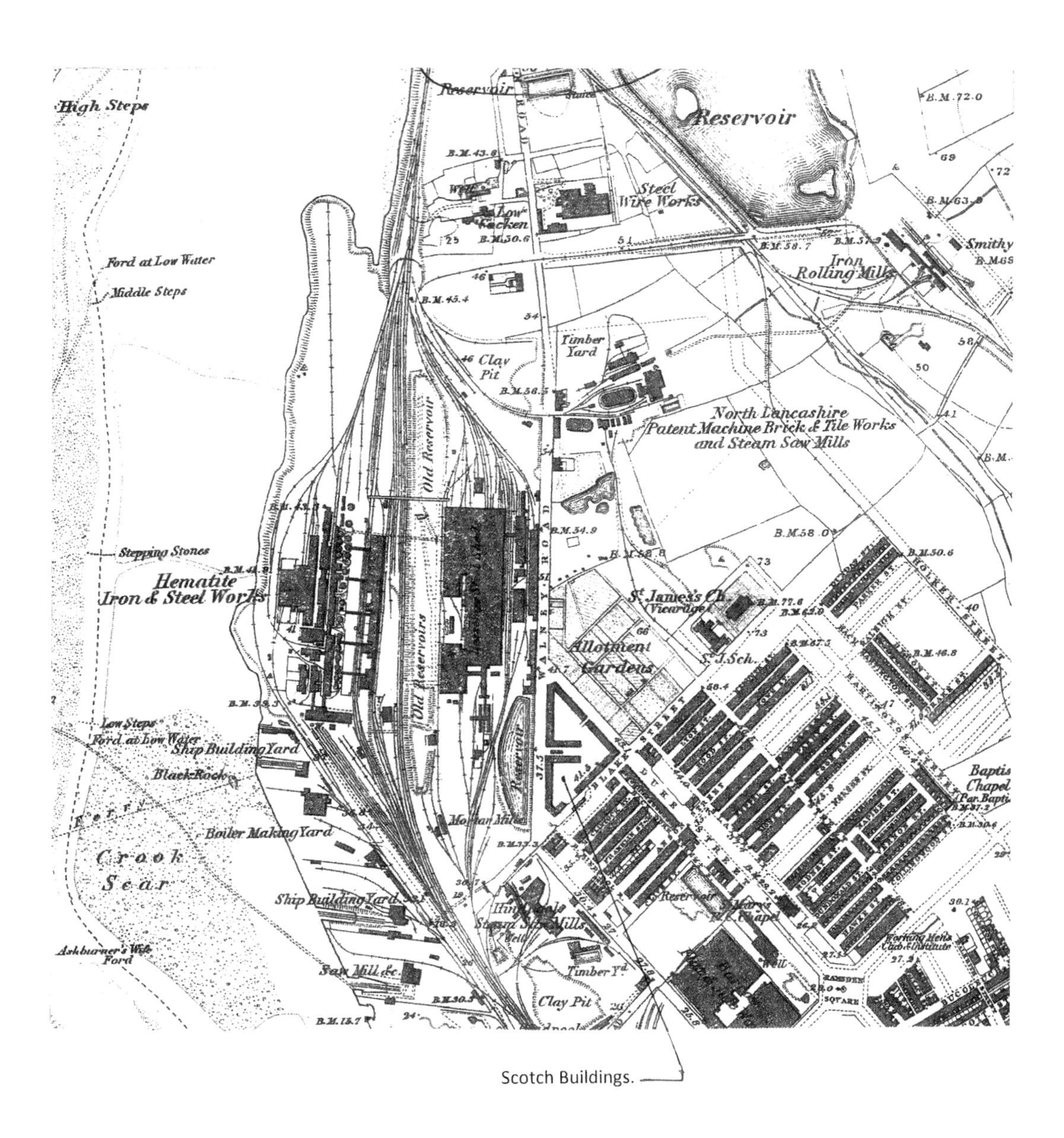

Map of Hindpool, extract from O.S. Map sheet 21, 1874.

2

The Hindpool Blast Furnaces

As originally constructed, the blast furnaces were 45ft in height, being raised in 1871 to 62ft. They were open-topped, allowing the hot gases produced to escape to atmosphere. The boshes of the larger furnaces were 21ft and the smaller ones 17ft in diameter. Each furnace was fitted with six tuyères. The blast, heated partly by Cowper and partly by Gjers stoves, (later upgraded to Whitwell) was originally 900°F, raised in 1890 to 1,270°F. The hoists were inclined planes, and fourteen furnaces were fitted with six inclines, each with a separate pair of engines (referred to later) and winding drums driven by friction gear and fitted with steam brakes. A high-level platform allowed the charge to be conveyed to each furnace by men known as 'barrow-wheelers'. The average weekly output of the furnaces was gradually increased as mechanical improvements were made, from a capacity of 500 tons in 1875 to 600 tons in 1890 and to 720 tons in 1898. The Hindpool furnaces were more than successful.

The furnaces were modified by Josiah Smith, general manager, for collecting and utilising the waste gases. Later, the furnace tops were closed off by incorporating the bell-and-hopper system invented in 1850 by George Parry of Ebbw Vale. The gas take-off was just below the hopper at what was called the furnace throat, and from here large pipes called down-comers directed the gases to ground level. Initially these gases were directed back to the foot of the furnaces to heat the blasts. Further developments saw these gases distributed to the steam-raising plant around the works – recycling before the term had been coined! From the mid-1870s the Hindpool blast furnace plant was deemed the finest and most modern in the country.

Initially the coke for use in the blast furnaces was brought in from Durham, 130 miles away, by wagons of the Furness Railway. The costs associated with this were more than compensated for by the benefit of having the richest of haematite ore on the doorstep. Years later, when the local ores became depleted and the basic method of steelmaking became established, these carriage costs became significant.

One day in early March 1880 the charging gang at No. 2 hoist had finished loading their carriage with coke, which was to be hauled up the incline and barrowed to No.5 blast furnace. The engine driver, Bill Woodward, was given the signal and the four barrow-wheelers mounted the loaded car as it began its ascent. About halfway up they noticed their speed was faster than normal and, as they weren't slowing as they approached the top, prepared to jump clear. At the top all four men made a spring to gain the upper platform, but William Hartley, 46-year-old father of five never managed the leap and, as one of the winding ropes had broken, was now on a white-knuckle ride back down to the charging depot. The wrought-iron carriage smashed into the stops at the bottom, flinging Hartley violently forwards and trapping him between the barrows. He was removed to hospital where it was found his skull was fractured, his jaw broken, and his leg broken. He died the next day from 'severe concussion'. The *Barrow Herald* of 23 March 1880 carried a report of the inquest. Although very tragic there was overall a low number of fatal accidents at Barrow works, which at the time employed around three to four thousand men at Hindpool. The incident was deemed statistically insignificant.

After the First World War, when mechanical charging had become the norm, the blast furnaces were reduced to four in number (Nos 6, 8, 10 and 11). A Halberg-Beth gas-cleaning plant was installed. The gases for the steelworks were piped away in a mitred 36in-diameter pipeline, which became known as the zigzag main. This pipeline straddled the tract of land, a kind of no-man's-land that carried the Furness Railway goods line bisecting the works. In the late 1860s there had been reservoirs situated here and then later banks of coking ovens. From the 1890s it was just a mass of sidings alongside hundreds of tons of iron pigs. Just north of where No.12 blast furnace stood was a wrought-iron footbridge, 17ft above the ground, enabling safe passage between the two sections of the works. The waste-gas system contained five stages of filters that removed unwanted particles, and the process could handle up to 5 million cu. ft of gas per hour. In addition to steam-raising for the mill engines, the gas was used to generate all the electricity required by the works.

Casting was almost a continuous and labour-intensive process. To the east, sloping away from the furnaces, were the sand pig beds. During the 1930s the first of two pig casting machines was constructed to the south of the pig beds, these were more compatible with the larger-capacity blast furnaces. Iron was tapped from the furnaces into ladles, which sat in sunken sidings, and then transported by rail to the casting machine. When in position at the base of the machine the ladle was tilted and the molten iron flowed along a short channel to the foot of the double row of mould pans. The moulds were attached to each other in a long chain and the chain moved along like a conveyer belt. As they filled, the line of pans moved up an incline; by the time they had reached the top the iron had solidified and, as the conveyer turned under to return, the pigs fell out of the mould, through a spray of cooling water, into a waiting railway wagon.

On the Walney Channel side of the furnaces were the hot blast stoves. These tall cylindrical structures stood to the west of the furnaces and were where the blast-furnace gas

View looking east from Walney Island. To the left is the machine shop, hot-blast stoves extend across the centre of the photo and the gas engine house is on the right. (Courtesy of F. Strike)

Aerial view of the works in the 1930s with the Scotch Buildings visible at top right. (Courtesy of Ken Norman)

was collected and burnt. The heat produced was absorbed by a honeycomb of special bricks in each stove, and then stored until needed for the next blast. Additionally to the west was an array of service workshops including a brass foundry and boiler shop to support and facilitate production. The post-war furnaces were each capable of an output of over 2,000 tons per week. At least two furnaces were in blast at one time, and in 1949 a record of 5,340 tons of pig iron was produced in one week. The bulk of the iron ore at this time was imported from Scandinavia, Spain and North Africa, augmented by supplies from West Cumberland.

Contrasting the previous image, the works viewed at night across Walney Channel in the early 1960s, the glow to the right being from the pig casting machine. (Courtesy of F. Strike)

Casting at night in the early 1930s, direct from furnace into the sand pig beds.

A view looking north across the sand pig beds *c.* 1930, with the zigzag main crossing to the right. Note the gas down-comers hugging the boiler-plated sides of the blast furnace.

The Smelting Process

The ore, coke and limestone formed the furnace charge and was carried up the inclined planes from the charging depots alongside by the hoist engines and dropped into the furnaces. Air was blown into the base of the furnace through the *tuyères* (pronounced twee-airs), which caused the coke to burn fiercely. Heat and gas were generated. The coke burned and sank and the gas flowed upwards. The gas and heat acted on the ore and together caused it to melt. The iron sank downwards in molten droplets and trickled into the hearth of the furnace. The earthy matter in the ore, together with any impurities, was taken out by the limestone and formed a slag, which floated on top of the iron. Slag and iron were tapped from the furnace at different levels, the iron being tapped every 5 hours or so, and the slag every two. The raw materials – furnace burden – were charged continually by way of the inclined planes mentioned earlier. Not only did we get iron from the blast furnaces but also several by-products, namely gas and slag, which could be further processed into building materials and ballast, as well as fertiliser for agricultural use.

The ratios: to make one ton of haematite iron required 1.8 tons of haematite ore, 1 ton of coke, 9cwt of limestone and 5½ tons of air. The rich and pure nature of the local ore allowed the furnaces to be tapped more frequently than with other, lower grade ores.

Preparing the foundations for a new blast furnace *c.* 1923.

Looking west and another view of the zigzag main. (Courtesy of F. Strike)

A 30-ton-capacity ladle on its car as used for transporting molten iron from blast furnaces to the metal mixer or the pig casting machines. The ladle and car would sit in a sunken siding alongside the retaining wall of the sand pig beds. (Courtesy of Richard Byers)

Teeming molten iron from the ladle onto one of the pig casting machines, 1962. (Courtesy of F. Strike)

View from the ladle end of the pig casting machine. (Courtesy of F. Strike)

View looking west in 1960. The two pig casters can be seen to the left amid the water vapour.

Plant from the modern era: device for plugging the tap-hole at the foot of a new furnace, 1960. (Courtesy of F. Strike)

Tapping a blast furnace manually. The tapping bar is just visible in the furnaceman's hand, and the stream of molten iron coming from the tap hole was referred to as 'the runner'.

The 1950s was a decade of modernisation, concluding in the early 1960s with sudden and unexpected closure. With the full co-operation of the British Transport Commission an ore-handling facility had been installed at Barrow docks so that iron ore could be discharged from ships' holds direct into receiving bins for onward transmission to either Barrow or Millom ironworks, and during 1956 a record tonnage of ore was received into the docks. Meanwhile, at Barrow a new Dwight-Lloyd sinter plant of 11,000 tons capacity had been installed together with a totally rebuilt blast furnace. In addition to iron smelting – and Barrow had carved something of a niche with its semi-phosphoric pig iron – production engineering facilities were expanded through an arrangement with an American manufacturer allowing the company to offer extended ranges of blast furnace and steelworks plant, specialised valves and other equipment. This arm of the business traded as Barrow-Kinney.

With the demolition of all the old steam-raising plant at the steelworks site in the late 1950s, its energy requirements now precluded the need for blast furnace gas, which for years had crossed the London, Midland and Scottish (LMS) Railway (formerly the Furness Railway) goods line via the zigzag main. Gone was the Monday morning ritual of Ken Royall meeting with Jim McWhan to discuss gas-meter readings and negotiate tariffs. The resourceful Mr McWhan had successfully negotiated an arrangement with

A view of the extensive engineering facility *c.* 1960. (Courtesy of F. Strike)

the British Cellophane factory at Sandscale. This factory, owned by the Cortaulds group, was set up on the outskirts of the town in 1959. Following the satisfactory agreement a 36in-diameter pipeline was fabricated and laid between the two works alongside the northbound railway line out of Barrow.

The *Barrow News* of 18 January 1963 reported that union officials representing the workforce had been told on the 16th that the works would close on 31 March 1963. Barrow works had been bought by Millom Haematite Iron Company. Millom ironworks were a part of the Cranleigh group and had secured the sale for £1,500.00 with the Iron and Steel Holdings and Realisation Agency.

The news that broke on that day in January 1963 was devastating for the town. The Ironworks was the town's oldest industry. Over 700 men would become unemployed. The reason cited for the closure was apparently an over production of pig iron that was building up throughout the Hindpool works.

In the meantime James McWhan had ideas for the engineering facility that had always complimented the blast furnace plant. He put forward a very strong case for retaining the large machine shop and running it as a 'stand-alone' business. His idea caught the attention of the Duple Bus Company of Blackpool who had the confidence to make the necessary investment. The initiative became known as the Barrow Engineering Company, which grew from a small focus to employing around sixty people. Jim McWhan stayed on in the capacity of managing director and the business traded successfully until 1984.

View west in 1963 across the works at the now silent blast furnaces, by which time that umbilical cord known as the zigzag main had gone. (Courtesy of F. Strike)

Following closure the works stood derelict throughout most of 1963 while officials scoured the country for potential buyers of items of plant still in working order. The *North-Western Evening Mail* of 2 November 1963 reported that items of plant were being dismantled and taken to the West Bromwich Company of C.C. Cooper, who were part of the Bromford Iron and Steel Company.

Among the plant purchased were three blast furnaces, eight water-tube boilers, gas-cleaning plant, turbo blowers, overhead cranes, water-mixing plant, pig casting machines and rolling stock. One of the blast furnaces had never been blown-in (used). Mr J. Scott, director at Barrow, said that the Midlands firm would be engaged on the site for another year, but the dismantling work should be completed before the end of 1964.

James McWhan, commercial manager. (Courtesy of F. Strike)

Looking east, the sinter plant dominates the view. Crossing the picture horizontally is the new gas main supplying the Cellophane factory at Sandscale. Stacks of unsold pigs can be seen in the foreground, 1963. (Courtesy of *North-Western Evening Mail*)

Demolition of a blast furnace under way. Examination of its firebrick lining reveals that it was never used, 1964. (Courtesy of *North-Western Evening Mail*)

The iron and steelworks site viewed from aloft in 1971. Note the blast furnaces have gone, and all that remains as testament to their existence is 9 million tons of slag.

How it was in 1871, from an original watercolour by George Henry Andrews, an industrial artist commissioned to paint several views of the works.

3

The Birth of Bulk Steelmaking

The Bessemer Plant: Final Remodelling

The ironworks of Schneider and Hannay were already up and running in 1863 when His Grace the 7th Duke of Devonshire visited Sheffield and witnessed an early Bessemer converter in action at Brown's Atlas steelworks. Within twelve months, following a successful pilot operation, work had started on the erection of a Bessemer steel plant at Hindpool.

By 1873 Sheffield had a Bessemer steelmaking capacity of a quarter of a million tons per year deployed in rail making and a further 100,000 tons directed at ships' plate. Even so, the Yorkshire city did not have things all its own way, since the 'Cumbrian' ironmakers had bought in.

The Workington Haematite Iron Company, with Sir Henry its principal stockholder, was part of the trend but by no means the main player. At Barrow a new holder of the world's largest steelworks title emerged, with fourteen blast furnaces and eighteen converters. The world steel capacity had doubled three times in the single decade of the 1870s and the two towns of Barrow and Sheffield were each making more steel at the end of it than the whole world had done ten years earlier.

At this point it seems fitting to paraphrase the words printed in the satirical magazine *Punch* on 5 October 1867: 'Barrow-in-Furness: from a barrow to a coach and four in ten years'. The skit was obviously aimed at Schneider and his burgeoning iron and steel empire.

Originally almost all the activity at Barrow steelworks was carried out in three large sheds with barrel roofs, built side by side and numbered 1, 2 and 3 from east to west. These buildings, each 700ft long and 110ft wide, were completed in 1865 and lay parallel to Walney Road. Once inside there was common access via bricked arches 16ft wide. Parts of 1 and 3 sheds were open to the outside, which ensured adequate ventilation for those working inside. The gable ends of each building were of local brickwork provided

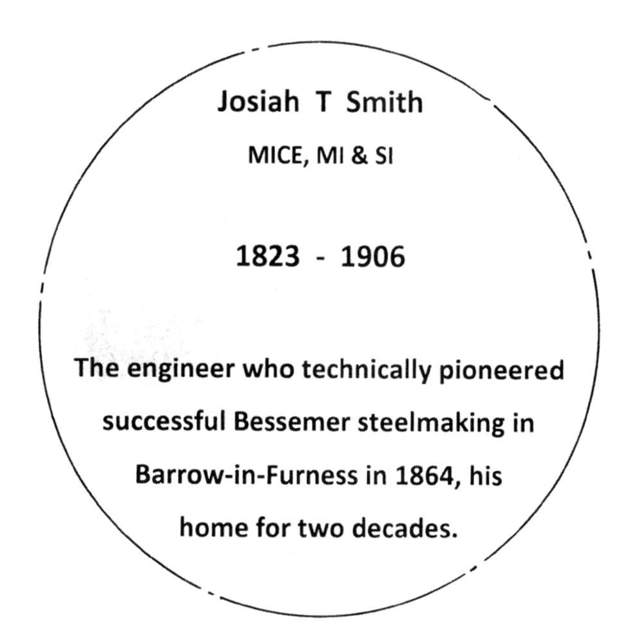

Transcript from the roundel at the front entrance to Chetwynd School, formerly Crosslands House, Rating Lane, Barrow.

Mr J.T. Smith.

Sir Henry Bessemer.

Mr J.M. While, M.Inst.C.E. General manager, moderniser. (Courtesy of *North-Western Evening Mail*)

with pitch pine windows at the upper segment. No. 2 shed had, at its apex in the southern wall, a huge bell used for signalling the start to work of the mills. The Bessemer plant occupied the northern ends of the three steel sheds and the remodelling was confined to No. 1 shed.

When the Barrow works were first established, the Bessemer process was in its infancy and the knowledge of the conditions best adapted to its economical application was limited. It was thought by Mr Smith, the first manager, that smaller converters would prove the most convenient and suitable, and hence the original plant was started in 1865 with a combination of 7- and 5-ton vessels – a greater number than has ever been built at any single works. The first steel made was on 23 May 1865. Ingot casting was achieved by the use of a piece of equipment referred to as a 'get' or 'git', which was basically a sand-lined conduit, so constructed to allow four ingot moulds to be filled simultaneously. The arrangement wasn't a new one; several firms had tried it but with little success. The casting pits were circular, with the ingot moulds placed in an arc within the swing of the transfer cranes.

In 1879 J.T. Smith remodelled the plant and the number of converters was reduced to eleven, of somewhat larger capacity. A few years later still, the Bessemer plant was again reorganised and the number of converters reduced to eight; and finally, in 1898 (just following the death in London of the inventor), the plant was entirely reconstructed in accordance with Mr While's up-to-date requirements and the number of units reduced to four.

All the iron used in the works was previously passed through the metal mixer – another J.M. While initiative. By this means, uniformity of quality was assured.

Barrow works was one of the first in Great Britain to adopt this important feature of the steelmaking process. Works without the benefit of a metal mixer had to rock the converter vessel to and fro when it had been charged with molten iron, almost like a giant cocktail shaker. At Barrow the metal mixer was placed halfway between the blast furnaces and the Bessemer plant, and the molten iron was run from the blast furnaces in large ladles specially constructed for the purpose. The mixer was turned via a worm and pinion on a main shaft, the worm spindle being driven by a high-pressure vertical engine, the cylinders of which were 14in diameter by 18in stroke. The air blow for the plant was provided by the Blast Engine House, located to the east of the Bessemer shop. The building was just inside the works boundary wall on Walney Road opposite the junction with Bath Street. The exact location of the metal mixer is not known, but the Bessemer shop was a quarter of a mile (as the crow flies) from the blast furnaces and separated by the Furness Railway main line. In reality the journey for the ladles of molten iron was over 1 mile. This was not ideal, and hence the requirement for four cupolas in the Bessemer shop.

James Morgan While originated from Glamorganshire learning his trade in the steelworks of South Wales, after which he spent some time in Sheffield. Upon being appointed to general manager he set up home with his family at Whinsfield, on Cocken Road. Once *in situ* his remit was simple – improve the efficiency of Barrow's processes. Starting with the Bessemer plant, he and Director W.F. Egerton, went to the USA on a study tour. During the tour he had been so impressed with the advantages of raised converters that, upon his return, he set about updating the Barrow plant. The Bessemer process comprised four steps:

1. Charging the vessel with molten iron.
2. The blow (about 15–20 minutes). After the blow a measured amount of molten spiegel was added to balance the carbon and manganese content, depending on the finished quality of steel required. (The spiegel was melted in the adjacent cupolas.)
3. Teeming into waiting ladles.
4. Slagging – emptying the slag into tubs.

The first step – charging – was always the potentially messier. Pouring the molten metal into the converter mouth could result in spillages. The converters were always charged in the horizontal position to minimise the length of time the molten iron was in contact with the converter bottom. Mr While therefore raised his converters some 16ft above ground level, then erected a staging to take the ladle cars. To the north of the Bessemer shop a long ramp was constructed of earth and ballast, creating an easy gradient up which the ladle cars could be moved by locomotives. This ramp (which cut right through where the *Evening Mail* building is on Hindpool Business Park) became known as Bessemer hill. The rearrangement allowed J.M. While's Bessemer output to match the original layout of Smith's but with up to sixty fewer men.

During the process of reconfiguring the Bessemer plant it was found necessary to remove the east section of No. 1 shed roof, which was in the way of the steelmaking activity. The building had originally been designed to accommodate much smaller converters and even these were set into pits below ground level. If Mr While had not had the roof section removed, the blast pillar from the 18-ton converters would have! Additionally, to protect the adjacent blast engine house a brickwork structure was erected, against which the 'venom' emitted from the converter vessels could impinge.

View of Nos 1 and 2 steel sheds from the south (rail bank end).

From about 1889 the USA had overtaken Britain in steel production and the basic method of steelmaking had become well established. Around this time the industry in Barrow began to suffer from lack of investment. Also, as good as J.T. Smith was at setting up the works initially, he failed to implement process control. In light of developments in the outside world, this would have allowed the business model to be changed. The Directors took handsome dividends, which of course was their right, but this reduced the amount of capital on the balance sheet to invest in plant and machinery. When J.M. While, who came with almost the same pedigree and background as Smith, took over he inherited a totally different scenario. He could update certain aspects of the works – subject to budgetary constraints – but could not influence market forces or the company's order book. What the works needed was another entrepreneur.

Workington steelworks, also struggling at this time, was later saved by being absorbed by the newly formed United Steel Companies of Sheffield – who were no doubt attracted by the West Cumberland reserves of ore and coal.

It would seem in another world that in 1893 an overexcited *Times* journalist was moved to write of a Bessemer in blow: 'A fountain of sparks arose, gorgeous as ten thousand rockets and fell with a beautiful curve, like the petals of some enormous flower. … Down came the vessel, until out of it streamed the smooth flow of terribly beautiful liquid steel.' A truly awesome sight to behold – Andrew Carnegie was probably right when he referred to steel as the eighth wonder of the world.

By the 1920s, when the supply of local ores was becoming exhausted, the demand for steel rails by developing countries had dropped significantly and the Belfast shipyard orders had been lost to cheaper American steel; this was keenly felt in Barrow. It is believed this factor coupled with the national coal strike, which caused steel production in this country to drop 55 per cent, convinced works management that Bessemer steelmaking at Barrow had run its course. No longer would the Hindpool works manufacture steel solely from haematite ore; its future would lie in scrap steel being recycled by the Siemens open-hearth process.

Bessemer converters are often mistakenly referred to as furnaces. They had no fuel supply or external heat input. Heat was generated chemically by the oxidation of the impurities in the molten charge – known as an exothermic reaction. Barrow's pilot converter (1864) is currently on display at the Science Museum, London under the heading of 'The birth of (bulk) steelmaking'.

The reference source for Barrow's metal mixer tells us that it was located mid way between the blast furnaces and the Bessemer shop. This would place it directly on the Furness Railway main goods line.

Mr While's Bessemer shop *c*. 1900. The wagon and workmen are on the raised platform. Note the inverted converter behind the wagon, while a converter mouth can be seen to the right of the group of men.

Not one of Barrow's but a new generation made in 1934. This particular model is on display at the Kelham Island Museum, Sheffield. Steel production by the Bessemer process in the UK ceased for good in 1974. (Courtesy of United Steel)

The above photo is of a 400-tons-capacity Wellman hot metal mixer being charged with molten iron, but where was the location of Barrow's? (Bessemer Steel Archive)

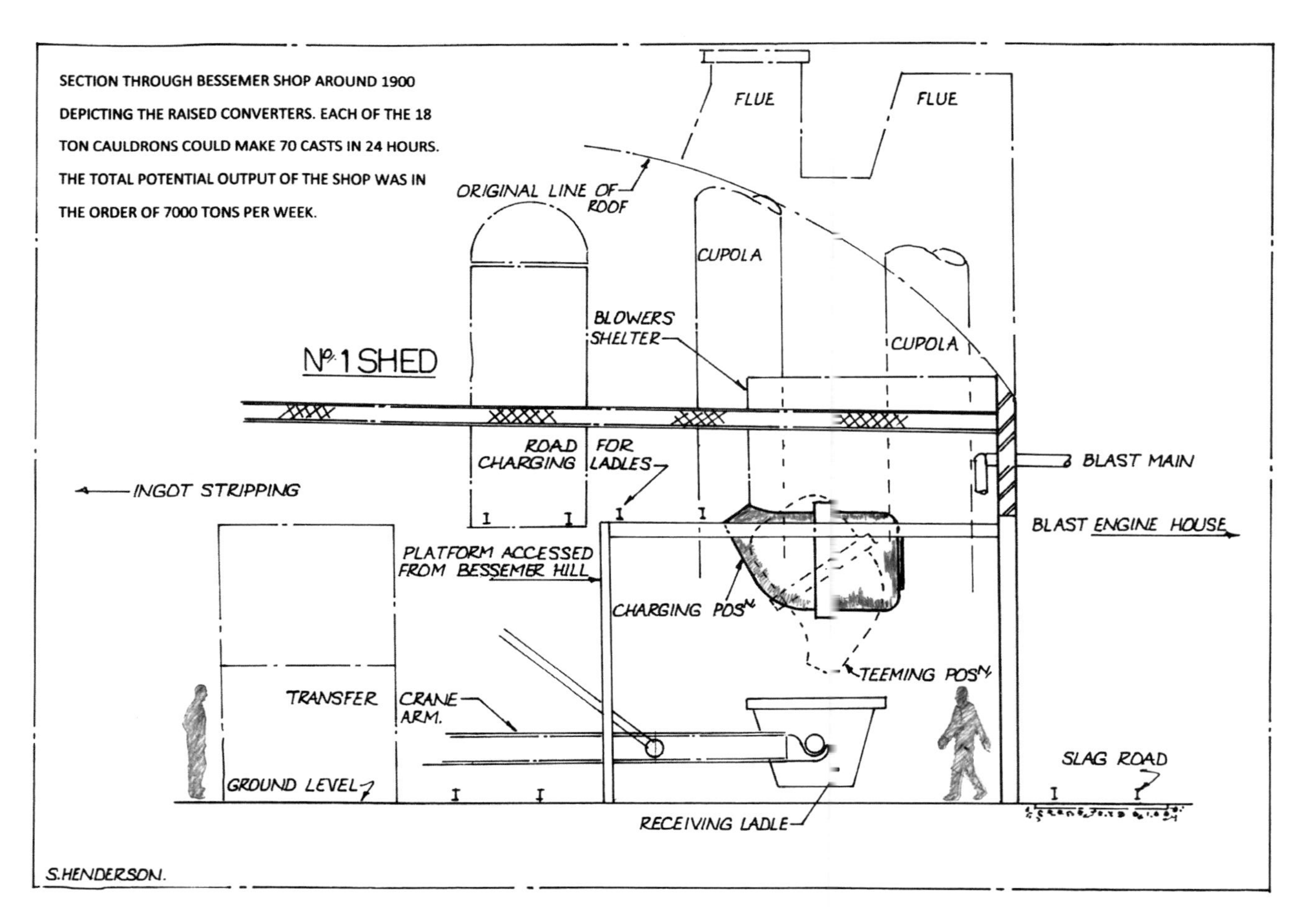

Section looking north through the Bessemer shop *c.* 1900.

4

Rolling Mills and Soaking Pits

The illustration below is from *c.* 1899 and shows No. 2 rail mill, with the cogging mill in the foreground. This cogging mill dispensed with all the steam hammers and their attendant commotion previously part of the original rail-producing activity. An ingot can be seen on the live rollers, which fed the mill, about to enter the second of the five passes. Ingots were brought from the soaking pits by an overhead crane called, because of its speed, the 'Whip'. To the left of the roll stands can

No. 2 rail mill, *c.* 1899. (Geoff Berry)

be seen the drive pinion that connected to the 2-to-1 gearbox and then into the glass engine house. To the upper right of the photo can be seen the underside of the 'chute'. This was a long banana-slide-type structure, unique to Barrow, up which extra-long rails going back and forth through the finishing mill would run, thereby keeping the mill floor clear. Ingots to be rolled down into rails were normally 2 tons in weight and 15in square.

The original mills, as laid down by Smith in the three steel sheds towards the end of 1865 and up to 1870, comprised a rail mill at the southern end of each shed, a merchant mill in No. 1 shed and a plate mill in No. 2 shed. Additionally there was an array of twelve steam hammers, supplied by John Musgrave of Bolton, of which two had 10-ton tups, sited across the three sheds. These hammers, in the pre-cogging-mill days, were used to reduce the cross-section of the ingots and involved several stages of reheating. By 1869 the three rail mills, which each were producing over 1,000 tons per week, struggled to meet demand.

The rail mills were of the three-high type. One of these was driven by a Ramsbottom reversing engine, the other two by beam engines, with 42in cylinders and 6ft stroke, supplied by Hicks of Bolton. The engines averaged 55lb boiler pressure and ran at 28 strokes per minute. The mills produced rails 100ft long and at the time were deemed to be of the first of their class. The original plate mill had two pairs of rolls 7ft 6in long and 28in diameter, with, on the other side, one pair 10ft long and 33in diameter. The mill was driven by two pairs of inverted-cylinder engines coupled, having four cylinders, each 26in diameter with 36in stroke, working at 70 to 100lb pressure.

An indication of the rail-producing capability at Barrow was given in the *Iron and Coal Trade Review* of 4 August 1899, which noted:

> The history of the Barrow Company is practically a record of the evolution of the modern rail mill, at least for a period of thirty years. The process of manufacturing rails at Barrow was similar, except in matters of mechanical detail, to that adopted at other Bessemer works. The ingot was brought from the converters, charged by a crane into the intermediate soaking pits, drawn from the furnace by another crane, and placed on the live rollers without any manual labour, except such as is needed to guide the piece from the last pass in one mill to the first pass in the next. Live rollers finally convey the rail to the finishing machinery, where it was straightened, drilled and inspected. Rails could be rolled at Barrow up to 250 feet in length and the mill was so arranged that almost unique facilities exist for the production of long rails. For this reason railway companies that required extra-long rails would give Barrow works a preference.
>
> Beyond the rail bank, which was of exceptionally large dimensions, new sheds were built for the purpose of affording shelter to the men engaged in measuring, punching and straightening.

The rail mill from the 1890s, sited in No. 2 shed, consisted of a 36in cogging mill, driven by a pair of 40x60in horizontal reversing engines, geared 2-to-1, with a 28in roughing

The men of the rail-finishing department, which was south of the rail bank *c.* 1900.

mill, having a pair of 48x54in direct-acting horizontal reversing engines, and a 26in finishing mill with similar engines. The three pairs of engines developed about 15,000hp. From the cogging mill to the cooling bank, the ingot would cover a distance of nearly 600ft, emerging as a finished rail. From ingot through to finished product the piece would retain its unique identity (cast number), ensuring full traceability.

Adjacent to the rail mill, and in the same building, was the merchant mill. This was a multi-purpose mill of the three-high type, which, as well as making fish-plates, colliery rails, angles and channels, also rolled a range of merchant bars and sections. This mill was closed down and dismantled around 1939/40 when the production of rail tracks ceased at Barrow. Orders for merchant bars from this time were passed to No.1 bar mill at the Hoop works.

Rail production by 1900 had more than halved, as Barrow's best customer – the USA – had overtaken Britain in steel production, thanks to a Scot called Andrew Carnegie (wee Andra'). For some time after their Civil War the US produced no rails from steel. Their early tracks were of iron and short-lived compared to the Barrow steel rail noted, even in the early days, for its good mechanical properties (Ultimate Tensile Strength (UTS) 30–32 tons per square inch). By the mid 1870s Barrow steel glistened across the North American continent, particularly throughout New York State where they were busy replacing their iron tracks with Barrow steel. Between 1867 and 1869 over 15,000 tons of rails had been shipped to the USA – and this was in the latter days of the Wild West! At one point the *New York Tribune* reported that 'it's now over three years since the steel rails from Barrow-in-Furness were laid, and are still showing little signs of wear'.

Two photos of an ingot being drawn from the soaking pit. (Images courtesy of Richard Byers)

The ingot (now a bloom) entering the last pass of the roughing mill, 1930s.

Despite tough times during the 1890s when the works closed several times, the company still had orders coming in for rails. In April 1898 the cargo ship *Samoa*, at 7,000 tons the largest to dock at Barrow, came in for mechanical repairs. She left Barrow six weeks later bound for Singapore with 8,000 tons of rails – obviously overladen. Then in May of the same year it was reported in the *Barrow Herald* that the works had signed a contract with the East India Railway for the supply of 9,000 tons of rails.

During the 1890s the plate mill had been modernised and resited in No. 1 shed. It consisted of a slabbing mill, furnished with rolls 75in long by 39in diameter, which reduced the 5-ton ingots to 30in wide and 4, 6, or 8in in thickness; a plate mill had two pairs of rolls, each 7ft 6in long by 28in diameter, the soft roll being of steel and the hard one of chilled cast iron. This mill was driven by a pair of Galloway reversing engines, with cylinders 50in diameter by 4ft 6in stroke, geared 2-to-1 and working at 80lb pressure. Each set of rolls had its full complement of live rollers and skids for the expeditious dealing with the rolled plate. An unusual practice of the plate-rolling process was the scattering of besom onto the white-hot plates as they entered the various passes. Besom was brought into the works daily by horse and cart. The heavy plate mill output was in

Entering the last pass of the finishing mill en route to the hot saw.

the region of 1,200 tons per week. Plate thicknesses produced ranged from ¾in to 1½in and these were supplied to shipyards in Belfast and on the Clyde. In the same building was a light plate mill, which rolled down to ⅛in thickness. The weekly output was around 350 tons.

The start of the Second World War marked the end of seventy-three years of manufacturing permanent-way products at Barrow. Many have said that but for the war the works would have closed. With the conflict came the demand for steel to support the war effort – steel for shell casings and angle frames for Anderson shelters were in demand. The government of the day saw the benefit of having steel manufacture, at Barrow and Workington, situated in remote areas and on the farthest coast from the enemy. Barrow works emerged from the war unscathed although Hindpool, for many years, bore the scars of bombing intended for both the shipyard and steelworks. During 1940 No. 2 rail mill was adapted for the rolling of billets for further processing in the re-rolling mills at the Hoop works. Additionally, steel in the form of large 7-ton ingots, as well as billets, were shipped out by rail to other steel mills in the UK.

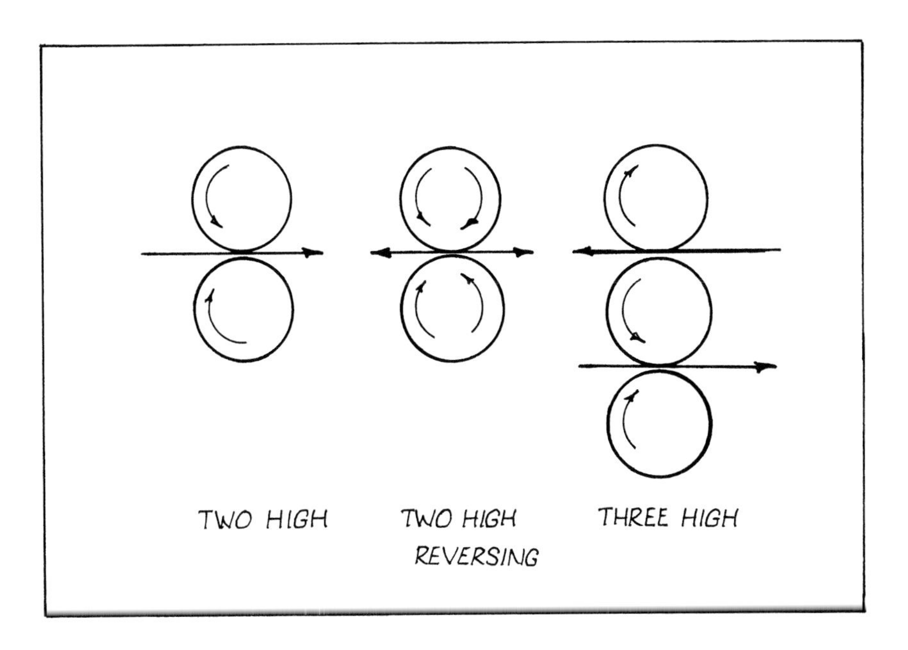

Diagram of the roll configurations used at Barrow.

The heavy plate mill in No. 1 shed. Although this is the actual mill, examination of the roll profiles at the time of the photograph show that the mill was not set up for rolling plate.

The slabbing mill in No. 1 shed in the 1890s. Note the large unguarded gear wheel to the right.

Sections of Barrow rails taken at the Colorado Rail Track Museum, Gold, USA in October 2002. (Courtesy of Les Gilpin)

Opposite page: Despite losing the US market, Barrow steel still had international appeal.

BARROW HÆMATITE STEEL CO. LTD.

BRANCH OFFICES AND AGENTS

London Office:
ST. MARGARET'S HOUSE, 57/59 Victoria St., London, S.W.1.
Telephone—Victoria 5246. *Telegrams*—Sillaw, Sowest, London

Midland and Yorkshire Representative:
J. N. WIGGINS, "Trevennel," Driffold, Sutton Coldfield.
Telephone—Sutton Coldfield 1468

Manchester Office (Lancashire, Cheshire and North Wales):
Chronicle Buildings, 74 Corporation Street, Manchester 4.
Telephone—BLAckfriars 1455

Scottish Agents:
T. RUSSELL & SON, 166 Buchanan Street, Glasgow, C.1.
Telephone—Douglas 1245 *Telegrams*—Accomplish, Glasgow

Continental Agents for Pig Iron:
WM. JACKS & CO., Winchester House, Old Broad Street, London, E.C.2.
Telephone—London Wall 7955
Telegrams—Alkalize, Stock, London

Canadian Agents:
HUGH RUSSEL & SONS LTD., 485 McGill Street, Montreal.
Cables—Russel, Montreal

South African Agent:
H. S. WILKINSON, 37-38 Standard Bank Chambers, Johannesburg, S.A.
Telephone—Central 7500
Cables—Prosteel, Johannesburg

Egyptian Agent:
G. CYRIL MORRIS, P.O. Box 403, 4 Bd. Saad Zaglul, Alexandria, Egypt.
Telephone—Alexandria 175
Cables—Sillaw, Alexandria

East African Agents:
LESLIE & ANDERSON LTD., 14 Billiter Street, London, E.C.2.

Also:

P.O. Box 150, Mombasa P.O. Box 324, Kampala
P.O. Box 32, Zanzibar

Two views of the interior of No. 1 steel shed but eighty years apart. The bottom picture (courtesy of Grace's Guides) shows a steel plate being sheared to size in 1897. The top picture is of the inside of the shed – the one-time home of the plate rolling mills – in 1977.

Post war and a view of the billet stock area to the south of the steel sheds.

A view south of the New Bank showing the billet-straightening and chipping area, while the Scotch Buildings on Walney Road can be seen to the right.

5

The Hoop and Bar Mills

There could have been no way of knowing in 1893, when the wireworks of Messrs Cooke and Swinnerton became just another acquisition of the Haematite Steel Company, that less than fifty years later, with only the odd exception, they would become the *raison d'être* for the steelworks' existence.

John Rushton. (Courtesy of Russell Rushton)

John Brimelow Rushton. (Courtesy of Russell Rushton)

John Maltby Black's family had a long association with the works, as far back as to being involved in the initial setting up of the factory. Even so it would be a Yorkshire man who would become the first works manager. John Rushton, from a coalmine-owning family, arrived in Barrow from Sheffield in 1871 to manage the new works that began operations in September of that year. The Rushtons settled in two of the new properties built on Raleigh Street, Hindpool and they employed two servant girls who had come down the coast from the Workington area. John was eventually succeeded as wireworks manager by his son, John Brimelow Rushton, who worked his way up from iron-turner to foreman, then manager. The wireworks comprised initially of two corrugated sheet buildings with floors covered in heavy iron plates. The works were situated on a 12-acre piece of triangular land bounded by Cocken Road, Devonshire Road and the FR loop line. The wire-drawing mill was in one building near to the Cocken Road boundary and contained twenty-four crucible furnaces for melting the steel off-cuts (rail ends and other arisings) supplied from the steelworks. There was also in this building fourteen drawing blocks for manufacturing the wire. The wire mill was driven by one of eight steam engines and produced a billet some 14ft in length. These billets would next be fed into a second mill driven by a larger engine, which turned the billets into wire via the wire drawers. One quarter of a mile of wire could be made from one billet and the works could turn out 14 tons of wire in twenty-four hours. The factory also made wire nails and bucket handles in large quantities. To the east of the wire-manufacturing department, in the second building, were the three-high hoop mills. This shed was 210ft long by 40ft wide. An extensive range off hoop and strip was rolled, and these could be supplied punched, nosed, or splayed and either varnished or oiled. Nine Cornish boilers supplied steam to the eight mill engines. Between 200 and 300 hands were employed.

In 1921, to cater for the expanding motor industry, a new facility was laid down called the Double-Duo. This was two bar mills side by side ('A side' and 'B side'), serviced by separate furnaces and manned by dedicated teams. The original mills were of the two-high type with 12in rolls in five housings. In later years 'A side' tended to specialise in rolling matchet steel, which Barrow works had helped to develop. Barrow had the recipe just right as the cutting edge of the tapered section had to satisfy exacting requirements. Upon leaving the finishing rolls, the matchet strips were placed in a jig called 'the box' to assist even cooling. From Barrow the strips were sent to a firm called Martindale's in the Midlands for finishing. The 'B side' mill, of similar proportions to A side, used to bolt down 4in by 2in slabs. These could be, for instance, silica-manganese for automobile and railway springs. The slabs had to be sufficiently and evenly heated otherwise damage to rolls could ensue.

In 1929 a young by-turn roller was loaned by the company to help with a rolling mill project in Spain. Many years later, the by-turn roller – Ernie Ward – was promoted to works manager.

This photograph is of a group of managers from the early 1900s, with John Brimelow Rushton on the back row, far right.

The Double-Duo looking north from the Stein furnaces, 1923. The suited man in the centre is J. Maltby Black, works manager and company stalwart.

The Double-Duo. Gavin Stamp and Ray Charnley are at the shears cutting matchet steel to length, 1967.

Looking from the steelworks across Walney Road to the Hoop works. Note the overturned billet-loading crane.

No. 1 mill cooling and inspection bank, 1936.

Batches of Y-section fencing standards from No. 1 mill awaiting shipment to Australia.

The older mills in production, considered in numerical order, included No. 1 bar mill. This mill produced a larger variety of sections than the others although it was not a fast mill as quality was more important than volume. The layout comprised four sets of stands of the three-high type. The mill was served by a gas-fired furnace, with gas supplied by a Sahlin gas producer. Sections rolled included Y-section fencing bars, T-section glazing bars and channels, a range of merchant bars to Admiralty specifications as well as bevel sections for helical springs. Finished bars were delivered onto an inclined cooling and inspection bank, from which rollers conveyed them to the finishing department.

No. 2 mill, a small jobbing-mill, rolled narrow strip 1in by 0.03in for cable tape and cotton ties. Strip issuing from the finishing rolls was discharged along a compressed-air-cooled channel 250ft long. From here it was inspected and tested for compliance with customer requirements prior to transferring to the finishing department.

No. 3 was a 17in mill and the main hoop mill. From 1952 it was progressively upgraded in parallel with the development of continuous casting to a cross-country, semi-automatic mill. It was only semi-automatic as it still needed an operative to manually transfer the piece from the plain rolls, around the mushroom, and into the first pass of the train of

Tongsmen in action on No. 2 mill *c.* 1936. The man on the left is employing a technique called 'looping' whereby he catches the leading end of the emerging strip as it leaves the rolls. He then turns 180° putting it into the next pass. In later years this manual task was achieved by the use of a 'repeater'.

A white-hot billet can be seen emerging from No. 3 mill furnace on its way to the bolting-down rolls on 31 October 1961.

The two 17in stands. On the far right is the entry side where the billet first passes through the de-scaler (water vapour) while in the centre of the picture hoop can be seen passing around one of the repeaters.

three finishing stands. It was a fast mill that broke production records and earned the mill hands a good wage. For many years it operated on a three-shift system. Billets for rolling were heated in a 'push' oil-fired furnace. Prior to automation three men per shift were needed for bolting down the 10ft long white-hot billets. These had to be drawn manually from the furnace, placed on a two-wheel bogey and wheeled up to the rolls. Such was the strenuous nature of the hot task that each of the three men took it in turn to rest for twenty minutes. The 10ft billets rolled out into 750ft of hoop.

Works engineer Frank Pearson was very much involved with the mill upgrade and regularly made liaison visits to the German works of the Muller Corporation who were providing most of the new components. By way of a reciprocal agreement one of the Muller engineers would come to Barrow and stay at Frank's home.

Unlike No. 2 mill the hoop coming off the finishing mill was directed to the coiling machinery after which it was bagged and labelled.

During the 1950s an extension was added to the existing warehouse to provide for the increasing output of the improved mill. This extended facility allowed for a larger finishing department and undercover loading of T. Brady's lorries – Brady's were the main haulage contractor to Barrow steelworks. On average Brady's would haul 600 tons of mill product per week from the warehouse – the combined output of 2 and 3 mills.

Left: exit side of the bolting-down rolls. The billet is just emerging, making its way along the 'wall' towards the stop.

On the right the billet, after hitting the stop, is dropping into the trough where rollers in the floor take it into the second pass of the mill, 31 October 1961.

The layout of the Hoop and Wireworks depicts the early 1920s when the new bar mill – the Double-Duo – had just been laid down. The new mills were driven by Siemens electric motors whilst the older mills were still steam-driven. In the not too distant future the existence of the entire steelworks would become reliant on the output of these small mills.

To the top-left of the Hoop works plan, accessed from Walney Road, was Cocken Villas. This house comprised two separate residences. During the 1920s J. Danks, Ironworks manager occupied one half of the villa while Siemens department manager H. Roberts lived in the other half. About thirty years later the building became the personnel services offices.

It can be stated here that the hoop mills at Barrow were the first to roll steel hoops, offering them for sale in 1872, and they were considered unsurpassed for ductility and uniformity of temper.

Note: The Steel Wire Works lost its corporate identity after being taken over by the BHS Co. in 1893, after this date it was just a department of the steelworks but to many it would always be the Wireworks or Hoop works.

The three sets of finishing stands. To the left is the tongsman who loops the piece from the plain rolls to the first pass of the finishing mill, 1961.

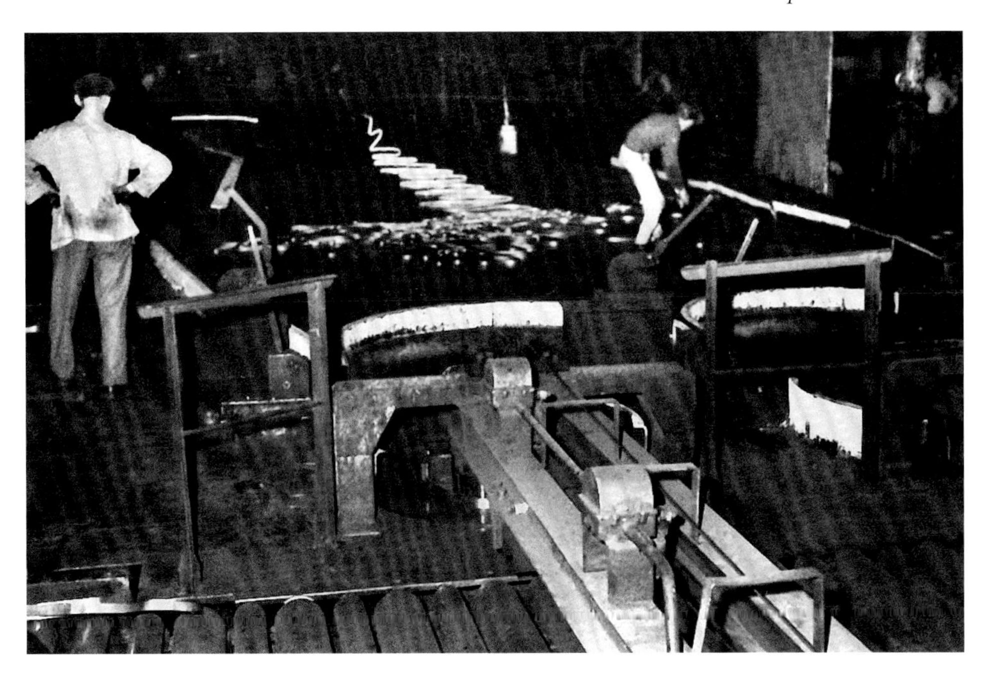

Red-hot hoop issuing from the finishing mill towards the coiling machinery in the 1960s.

No. 1 mill flywheel and motor, with foreman electrician Wilf Bethwaite looking on, 1978.

J. Maltby Black, retired manager, with his wife and daughter outside Buckingham Palace with his MBE in 1961. Maltby Black retired in 1954 after fifty-three-years' service. (Courtesy of *North-Western Evening Mail*)

Six of the women, and can-lad, of the finishing department, 1957. (Courtesy of Baines family)

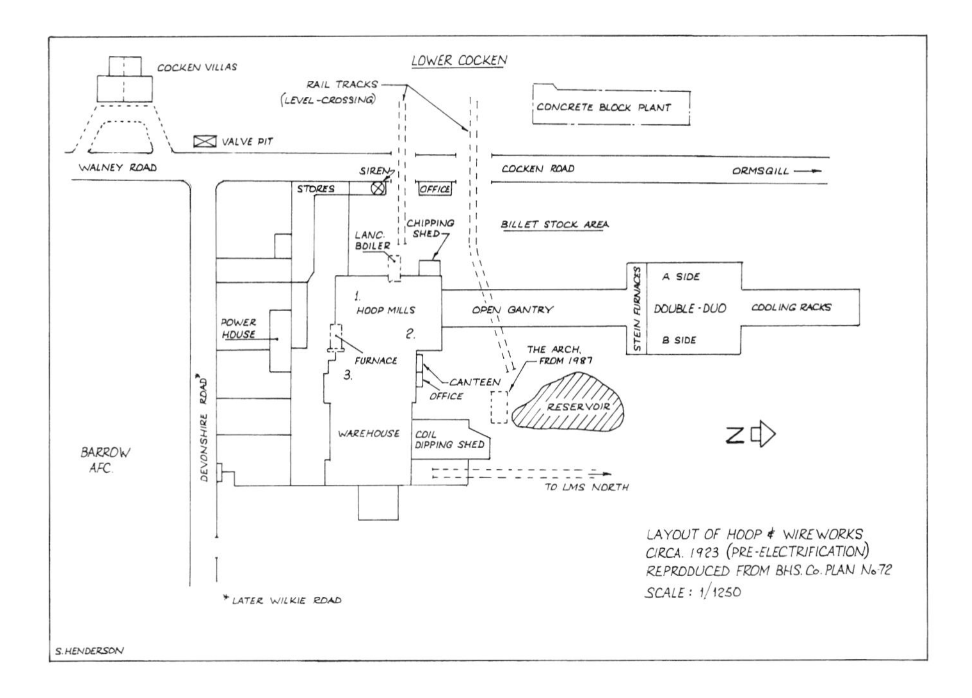

Plan of Hoop and Wireworks, 1920s.

BARROW STEEL has international impact — its customers stretch the world over. Rather more than 40 per cent of its mill product is exported.

Strapping goes to Sri Lanka and America; hoops to Spain; flats to Ireland and Singapore, machete to Africa.

A large part of its UK sales go to the whisky industry and North Sea oil rig construction.

Coopers Hoops

Whisky is matured in white oak wooden barrels. Strong hoops are required to compress the barrel and render it liquid proof for many years. Barrow produces a range of hoops, cut and splayed to the master cooper's exacting requirement. Barrel hoops are also supplied to the beer-making industry and for sherry casks.

Machete

Barrow produces the machete strip, cut to length, tapered and grooved for the final processing to be carried out by the customer. This is a high quality steel which must keep an edge but resist bending or breaking on impact. Barrow has helped to develop this steel, which with the skilled heat treatment of the final processors, meets these exacting requirements.

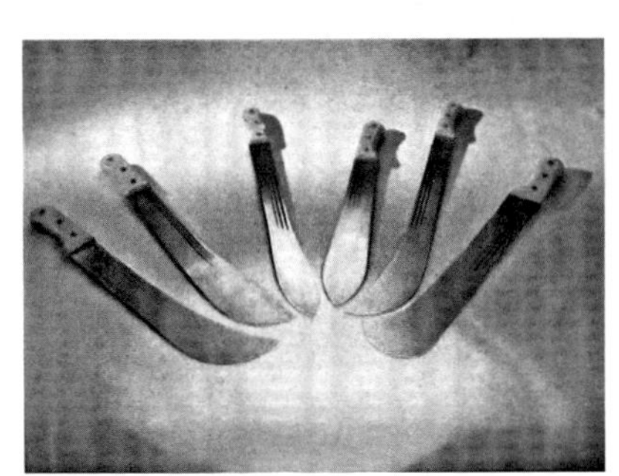

EXTRACT FROM BRITISH STEEL PAMPHLET

FEATURING HOOP WORKS PRODUCT RANGE.

Extract from British Steel pamphlet.

6

Open-Hearth Plant and Steel Foundries

The open-hearth steelmaking process was based on the work of William Siemens, a German engineer who later became a British subject. He came to the north of England twice, once as a young man and then later in 1874 when he spoke at the dinner given to the Iron and Steel Institute in Barrow. Siemens applied the regenerative principle to the furnace whereby a honeycomb of firebricks below the hearth absorbs heat from the waste gases and then returns it to incoming gases in an alternating cycle. This method could save up to 75 per cent of the fuel used previously. The open-hearth process allowed large amounts of scrap steel to be used, as well as pig iron.

The open-hearth plant at Barrow was launched on a modest scale compared to the Bessemer operation. This initiative allowed two things: it permitted a diversification of products, and it was one step towards making the steelmaking side of the works independent of the ironworks. In 1880 two 12-ton furnaces, with the necessary gas producers for six in case the plant was later extended, were laid down. These furnaces were arranged so that molten metal from either blast furnaces or converters could be used (Duplex system). Initially open-hearth metal was used to supply the three steel foundries, but during one particular week in 1881 the combined works output hit a record 4,360 tons – up to that time the largest quantity ever produced in one week by a single firm.

No. 2 foundry was 120ft long and 50ft wide, and an extension to the northern end of No. 3 shed. This foundry aligned the open-hearth casting pit and was referred to as the north-end casting pit. (Many years later it would become the home of the continuous casting pilot plant.) No. 1 foundry was 366ft long by 50ft wide and was located to the north of No. 2 shed. This brick building, from about 1930 became the boiler shop. No. 3 foundry was 250ft long by 80ft wide, situated to the south of the open-hearth plant, and would later become the Siemens laboratory and office. The largest steel castings up to 1920 were steel rolls, 28 tons in weight and stern frames of 22½ tons. The company were prepared to accept enquiries for castings of up to 100 tons, and castings had been provided for Belville boilers, gun carriages and all kinds of marine work. In addition to the three steel foundries there was also a foundry for making ingot moulds and slag ladles. This building, 150ft long by 60ft wide, was just inside the works boundary wall on Walney Road opposite the entrance to the North Lancashire brick works (currently an Asda superstore).

In the early days of the works ingots rarely exceeded 13–14cwt, being later increased to 35cwt. From the 1890s ingots averaging 5 tons were produced. (Around the time of the Second World War 7-ton ingots were made, but these were sold on for use in works with universal mills where large products such as joists were rolled.)

The open-hearth furnaces were increased in capacity over the years. The layout of the late 1890s comprised four 25-ton, four 50-ton and one 8-ton furnace. The total potential output was about 120,000 ingot tons per year. From the early 1920s the plant was again upgraded to three 80-ton, and four 40-ton furnaces (at least one of each size would be acid, the others basic – chemical composition is beyond the scope of this book although a brief

Marine casting in No. 1 steel foundry awaiting despatch, *c*. 1920.

discussion can be found in the appendices). Additionally, a Morgan gas-producing plant was installed at the external north-west corner of No. 3 shed. This plant fuelled the open-hearth furnaces as well as the two sets of klondyke soaking pits. These reheating furnaces, of the latest construction, sufficed to do all the work previously done by forty-six smaller furnaces supplied from seventy-two gas producers, sited to the south of the steel sheds.

At some point during the early 1920s steelmaking by the Bessemer process came to an end and the works began to buy in larger quantities of scrap steel for recycling in the Siemens furnaces. At the same time, across the LMS goods line, the ironworks continued to export its s.p. pig iron. This was an area where, following a close study of customer requirements, it was recognised that a percentage of phosphorus was desirable in irons used for the manufacture of certain castings. Research also revealed that castings with more than 0.35 per cent phosphorus frequently suffered from porosity and general weakness (a condition known as cold-shortness).

In the report of the 1880 visit to the works by the Institution of Mechanical Engineers, members reported witnessing ingot casting under steam pressure. During

A 5-ton Wellman charger emptying a batch of scrap into one of the furnaces.

The crew of F furnace, one of the 40-ton units. Left to right: Les Jones, leading hand John Thompson wearing white apron and sweat-cloth, unknown, unknown, 3 March 1953.

Alec Finch of the fuel department with a furnace-metal pyrometer, 1953.

the years collecting information for this book no internal record could be found covering this aspect of the process. One possible explanation is offered, but this needs further verification. It is known that on occasion dissolved gases in the liquid steel would 'effervesce' in the ingot mould and spill over – very similar to pouring out a bottle of beer too quickly. Stoppering-down after teeming with applied pressure may have been one way of countering this action. This may be one of the reasons why molten spiegel was added to the steel prior to casting. Spiegel contained manganese, which removed dissolved gases.

The larger open-hearth plant comprised static furnaces on a split level. The higher level was where the furnaces were managed and charged from an underslung Wellman charger. The amount of scrap steel and pig iron would be carefully weighed and recorded prior to putting it into the basket of the Wellman charger. The launders were on the lower level and were tapped by gravity into waiting ladles. These ladles sat on cars, which straddled the Siemens casting pit where rows of ingot moulds waited two abreast. Once the furnace had been tapped ingot casting would begin almost immediately.

The 40-ton-capacity furnaces were sited due north, under the same roof as the larger 80-ton units and steel was cast in the north-end casting pit.

The larger furnaces were later converted to oil whilst the smaller units remained on gas, but the mode of operation was the same. Air, heated by fuel and exhaust gases, was

Exterior view of Morgan gas producers, *c.* 1923.

Interior view of the Morgan gas producers showing the coal hoppers. These machines made the gas that fuelled the open-hearth furnaces and soaking pits, *c.* 1923.

Teeming from an 80-ton-capacity bottom-pour ladle at the pit side, slag is cascading into adjacent tubs. Note the suspended steel plate shielding the teemer and pitman from the intense heat, 1950s. (Courtesy of Bill Myers)

fed through ports at either ends of the furnaces with the direction of flow alternated every half-hour or so to produce intense heat. When the furnace charge was molten a sample was taken and sent for analysis to determine its composition. If necessary, fettlings would be added followed by another sample, a final bath temperature would be taken and then the furnace would be tapped.

The time between charging with scrap steel and tapping could be over seven hours. The tap hole would be filled with anthracite prior to charging. When the heat was ready the bung would be rodded out with hammer and crow bar, allowing the liquid steel to flow into the waiting ladle.

Steelmaking by the open-hearth process continued until late 1959, and the weekly capacity of the melting shops was about 3,500 tons. Approximately 600 to 1,000 tons went to the re-rolling mills weekly, the remainder being sold as ingots and billets. During 1960 the ageing furnaces, soaking pits, cogging and billet mills were totally demolished.

Another view of ingot casting from the 80-ton ladle.

Taking a metal sample from an open hearth via the working door.

Ingot casting into 2-ton ingot moulds in the north-end casting pit.

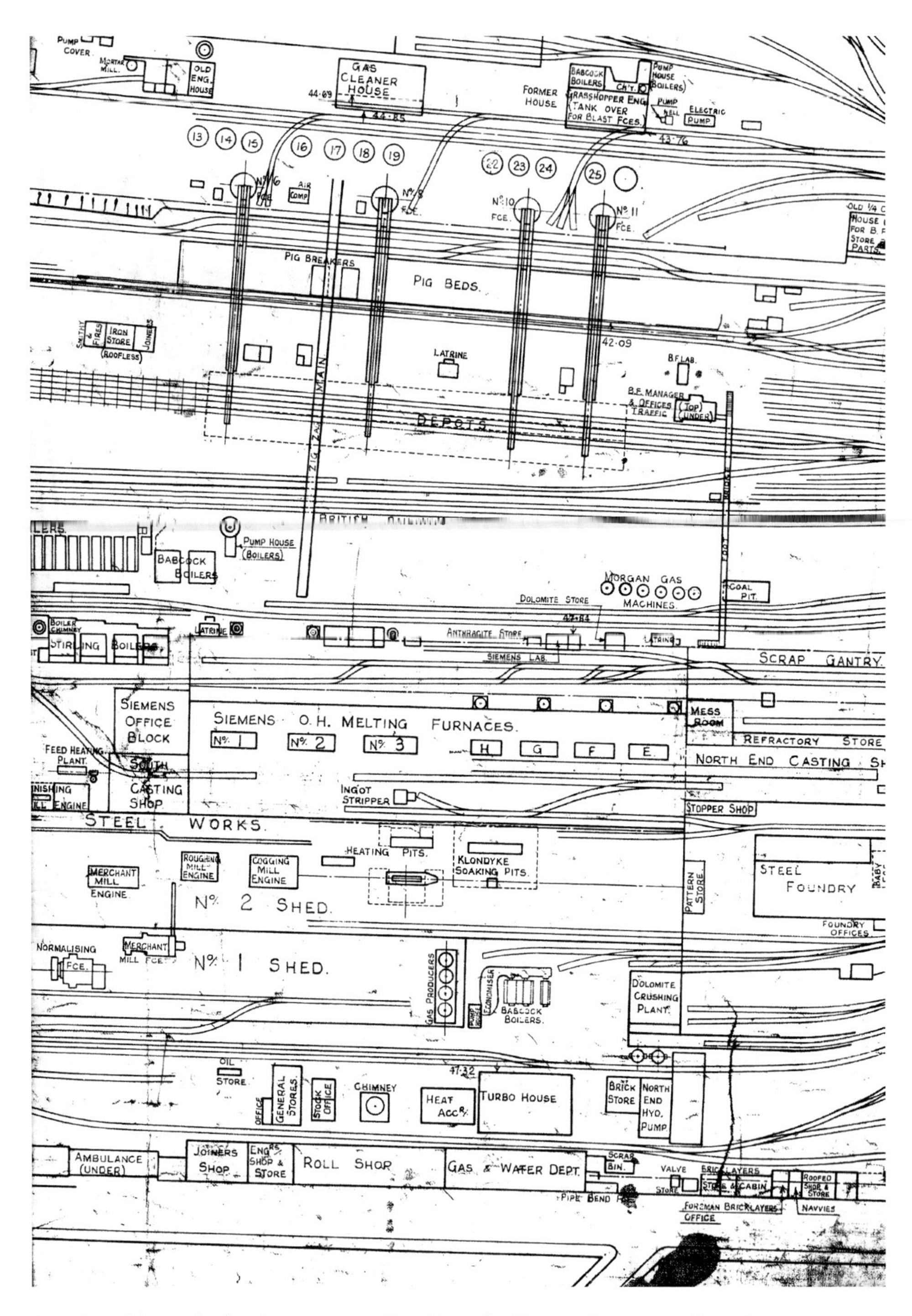

Part plan of the works showing, amongst other things, the Siemens furnaces, soaking pits, Morgan gas producers and steel foundries, reproduced from BHS Co. plan no. 72.

7

Boilers, Engines and Transport

To provide for such a large establishment, the boiler plant was on an especially large scale. As far back as 1880, the works had 150 boilers of all types, of which twenty-one were built on the Howard patent and fired by Vicars' mechanical system. A number of these were small boilers and to a certain extent were displaced, the number in 1899 being 160. Of those, the greater part were on the Cornish principle with dimensions of 33ft long by 7½ft diameter. Cornish boilers were developed around 1810 and featured a single fire tube running centrally along the length of the boiler surrounded by water. From about 1920 the Cornish boilers were replaced with batteries of Lancashire boilers, which were developed in 1844 by William Fairbairn. Lancashires had twin furnace tubes, side by side, which gave them a much larger heating surface than Cornish boilers. The addition of Galloway tubes, patented in 1848, brought a further improvement in terms of efficiency.

As part of the post-war modernisations (from 1942 onwards), the Lancashire boilers were demolished and replaced with three of the more efficient marine boilers. These boilers were evolved from Lancashires but were much larger and efficient units. On a smaller scale, Babcock and Stirling boilers were used across the site. Over the years as the works downsized, the steam-raising plant was reduced pro rata until, with the advent of continuous casting, the only steam raised was for domestic and heating purposes.

In the 1890s the works as a whole had twenty-three blowing engines, seven winding engines, and ten mill engines. The beam engines used for blast purposes were the largest on the works, each of them having blowing cylinders 100in in diameter with a stroke of 9ft. The incline, or hoist engines, at the blast furnaces had cylinders 16in in diameter with a stroke of 2ft. The Bessemer blowing engines were six in number, four being horizontal and one pair vertical, and gave a blast of 25psi. Most of them were made by Perry and Sons of Bilston. The original rail mill beam engines were from Hicks of Bolton. Apropos the blast furnaces, when the works were founded there was a considerable difference of opinion among authorities as to whether one or two engines of the larger size or a number of small engines would be best adopted for supplying the blast. The advisors to the company recommended the adoption of a number of smaller engines and hence the erection of no fewer than sixteen grasshopper blast engines, in addition to three beam engines already fitted up, making a total of nineteen blast engines. For many years the engine houses at Barrow were among the wonders of 'modern' furnace plants. In summary there were eight engines fitted up in one engine house, with room for two others, and an incline engine besides at each end. It was reported in the trade press that Barrow had the largest engine house in the world.

The gas engine plant (from 1909) comprised six by 1,850bhp double-acting, single-crank engines driving blowers and generators. This gas engine house was situated immediately inside the ironworks gate at the top of the (old) Ironworks Road.

The works were serviced by twenty-four locomotives, of which fourteen were standard and the rest narrow gauge. The rolling stock – post Second World War on the steelworks side – painted Ministry of Supply blue, was maintained by Harold McCracken who was on permanent secondment from the Wigan Wagon Company. Mobile steam cranes were used across the site. These were supplied by Jeremiah Booth & Co. of Leeds and ran along the internal rail network. This was maintained by the platelayers, whose foreman was Joseph Baker. Joseph had been at the works from his youth and, apart from a spell with British Rail in the 1940s, was a long-serving and experienced hand. It was said that when it was intended to fit a new junction at some point in the works, Joe would have it figured in his lunch hour. It would then invariably be installed in advance of the detailed drawing being issued by the works engineer's department.

Marine boiler house in 1961 awaiting demolition.

A view of the gas engine plant, 1909.

No. 2 roughing mill engines awaiting dismantling on 2 September 1959.

The interior of the turbo generator house, 24 May 1954. Originally the building housed the Bessemer blowing engines.

The roughing mill was driven by two 48in by 54in direct-acting horizontal reversing engines, geared 2-to-1. The gearbox can be seen between the engine and the roll stands. This 28in mill was of the two-high type and the live rollers for handling the throughput of either rails or billets can be seen in the floor either side of the stands. The dwarf wall, which was the base of the original glass engine house, can be seen with the driver's cab straddling the reversing gear, supported on four legs. When the engines were running at high revs the cabin would shake like a 'nervous wreck'. At centre, top of photograph, is the steam supply pipe and shut-off valve.

Throughout most of the 1950s the roughing mill took on the duty of the cogging mill. This was because the latter piece of 'ancient' machinery had been repaired so many times, it had to be permanently retired.

BSW loco. No. 7 was originally built for the Furness Railway Company as FR 20 by Sharp Stewart. This, along with six others, was sold to Barrow steelworks in 1890. They were modified into saddle-tank engines and used for shunting. In 1960 two of the engines (Nos 7 and 17) were gifted to school children: No. 7 was sent to the George Hastwell Centre on Abbey Road, Barrow while No. 17 went to Stone Cross Boys' school at Ulverston. The engines were cleaned and transported free of charge, and it was believed to have been a first for this type of donation. A few years later No. 7 was finally relocated to Haverthwaite (Lakeside & Haverthwaite Railway) after undergoing restoration to its original condition as FR 20 (with acknowledgements to *The Great Survivor* by Tim Owen).

Two photographs of No. 7 engine being loaded onto a Pickfords low loader for removal to the George Hastwell Centre in September 1960. In the top photograph can be seen the Siemens crane gantry.

The works' limousine was a dark green Austin that had leopard-skin seat covers and a voice pipe that connected the passengers to the driver in his open compartment. The company chauffer in 1920 was Joe Thistlethwaite. The bottom photograph on p. 81 shows the vehicle outside the company's guest house on the corner of Abbey Road and Dane Avenue, Barrow. This building has subsequently been the Glen Garth Hotel and is currently used as a care home. If, on occasion, the number of guests exceeded the capacity of the fourteen-bedroom building, the company used Furness Abbey Hotel – the residence of general manager Mr P. List.

The hotel was originally The Manor House at Furness Abbey and was converted to a hotel in 1846 from the opening of the Furness Railway, which had directors and associates in common with the Steel Company. The building was partly demolished after the Second World War, but the north wing – to the right of the photo – was left standing and for many years traded as the Abbey Tavern.

Loco No. 7 being enjoyed by the pupils of the George Hastwell Centre in September 1960.

Loco No. 7 resplendent again as FR 20 at Haverthwaite railway station, April 1997.

End of an era: demolition of engine 8/29, built by Kitson and Co. of Leeds in 1890.

Brady's eight-wheeler Scammell driven by Jack Thompson, hauling 20 tons of spring steel flats out of the works' main entrance on Walney Road *c*. 1965. (Courtesy of T. Brady and Son Ltd)

The northern end of the 200-yard-long Siemens scrap gantry, 28 July 1970.

The 10-ton Goliath electro magnetic scrap-handling crane installed by British Steel in 1977 to replace the Siemens gantry. (Courtesy of *North-Western Evening Mail*)

The company guest house and limousine, 1920s.

Furness Abbey Hotel photographed just prior to demolition.

Junior Ambulance Corps with General Manager Mr P. List seated to left, outside Furness Abbey Hotel in the 1920s.

Plant Ledger Dept.
Engineers Office.

January, 1948.

GENERAL TRAFFIC AND LOCO SERVICE (STEELWORKS) DEPT. NO. 507.

Plant No.	Description.
507/1.	Buildings and building equipment, including foundations, walls, roofs, stairways, doors, windows, all fittings on buildings and offices.
2.	Drains and fittings relative thereto.
3.	Electric power, lighting, fixtures, cables and wiring.
4.	Departmental railway lines and permanent way.
5.	✓ No. 1. Locomotive (Peckett 1935) with two inside cylinders 16" x 22", 0 - 4 - 0, 7'6" wheelbase, 12'0" total height 4'8½" gauge.
6.	✓ No. 4 Locomotive (Peckett 1937) with two inside cylinders 16" x 22", 0 - 4 - 0, 7'6" wheelbase, 12'0" total height 4'8½" gauge.
7.	No. 5 Locomotive (Sharp Stewart 1890) with two inside cylinders 16" x 24", 0 - 4 - 0, 7'8½" wheelbase, 12'2½" total height, 4'8½" gauge.
9.	✓ No. 7 Locomotive (Sharp Stewart 1890) with two inside cylinders 16" x 24", 0 - 4 - 0, 7'8½", wheelbase, 12'3" total height, 4'8½" gauge.
10.	✓ No. 17 Locomotive (Sharpe Stewart 1890) with two inside cylinders 16" x 24", 0 - 4 - 0, 7'8½" wheelbase, 12'4¾" total height, 4'8½" gauge.
11.	No. 22 Locomotive (Sharp Stewart 1890) with two inside cylinders 16" x 24" 0 - 6 - 0, 10'9" wheelbase, 12'3" total height, 4'8½" gauge.
12.	No. 27 Locomotive (Kitson & Co. 1880) with two inside cylinders 17" x 26", 0 - 6 - 0, 10'9" wheelbase, 12'2½" total height, 4'8½" gauge.
13.	✓ No. 8/29 Locomotive (Kitson & Co. 1890) with two inside cylinders 16" x 26", 0 - 4 - 0, 7'6" wheelbase, 12'4" total height, 4'8½" gauge. No. 8 Boiler on No. 29 Frame.
14.	North end water tank and water columns, including piping, valves, fittings and foundation (1890).
15.	South end water tank and water columns, including piping, valves, fittings and foundations (1890).
16.	Iron and steel wagons for internal use.
17.	Iron and steel wagons for main line use.
19.	Wooden wagons for internal use.
20.	Wooden wagons for main line use.
21.	Tar tank wagon.
22.	60.ton railway weighbridge (Ross No. 15695. 1934). including fittings and foundation. (North end of Siemens Gantry).
23.	50 ton railway weighbridge (Pooley No. 503) 1919. including fittings and foundations. (North end of Traffic).
24.	30 ton railway weighbridge (Hind. 1890) including fittings and foundations (Rail Bank south end Traffic)
25.	50 ton railway weighbridge (Pooley No. 503. 1921) including fittings and foundations (Rail bank south end Traffic)
26.	Hunslet Locomotive No. 2405 ex. R.O.F. Kirby, June 1946 (0 - 4 - 0) R.O.F.7.
27.	
29.	
30.	
31.	

Copy from an original works document.

8

The Development of High-Speed Continuous Casting

It had always been the dream of steelmakers that steel could be made in a truly continuous process, and during the 1950s Iain Halliday and the team at Barrow moved that dream closer to reality.

In 1943 the United Steel Companies of Sheffield took over from the Barrow Haematite Steel Company and became managing agents. At that time the works consisted of two open-hearth melting shops and a steam-driven cogging and billet mill while the re-rolling mills comprised three small bar mills and two hoop mills. With the exception of the Ironworks and Hoop works, the main steelworks was a huge inefficient Victorian relic. Initial modernisations included the establishment of an instrument and fuel department (gone were the days of spit and see), changing the larger furnaces to oil and the introduction of hydraulic actuation to furnace doors.

The United Steel Companies were at the forefront of investment and innovation. Through heavy investment they had improved the efficiency of ironmaking by the use of sintering. In their Seraphim blast furnace the entire burden was sintered. Steelmaking was speeded up by the development of the Ajax furnace. Around 1950 they turned their attention to continuous casting. The development of this process was undertaken primarily to meet the local requirements of Barrow works, but in addition they were sowing seeds for the future. Steelmaking had become increasingly competitive and as the United Steel Companies (USC) comprised four major iron-and-steelworks in the UK, they knew a process that eliminated the ingot/primary mill route would mean great savings.

The traditional method of producing steel was to tap molten metal from the furnace into a ladle and then teem the liquid steel into static moulds. When the steel solidified, the mould would be stripped from the ingot. These could be 2, 3 or 5 tons in weight. The ingots were then transferred to a reheating furnace called a soaking pit and heated to over 1,000°C. From the soaking pit the ingots were conveyed to the primary mill where

The Scientist Iain M.D. Halliday combined ferro-metallurgical knowledge with a determination to succeed.

The Engineer Frank J. Pearson provided the engineering solutions in adapting the Junghans process for use with steel.

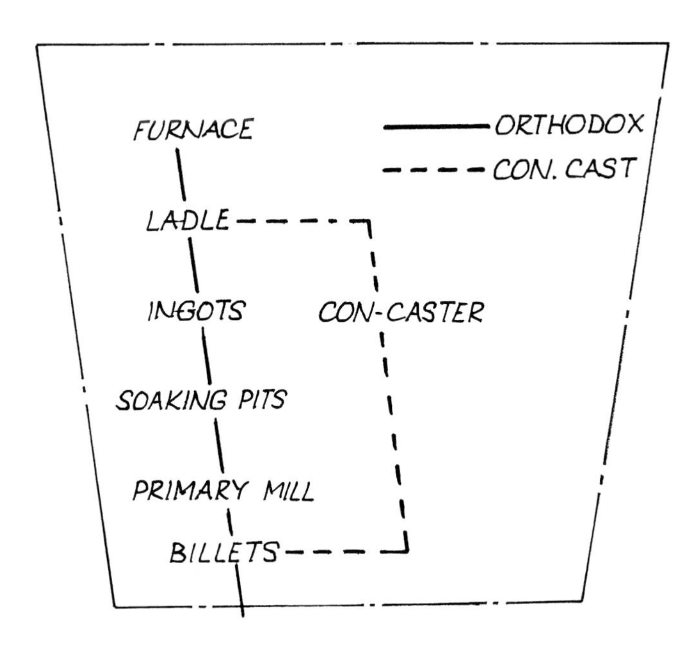

Key stages of steel production.

they would be cogged down to a smaller but longer size. Further rolling would produce a variety of different sections. Continuous casting would do away with the ingot, re-heating furnace and cogging-mill stages. The advantages to a steelmaker can be summed up by saying the process eliminates several steps on the route from molten steel to rolled product, uses less fuel, manpower and space and has a much higher yield of sound, usable steel.

Continuous casting is fundamentally simple. Seen in action it is deceptively so. Molten metal is poured at a controlled rate into a bottomless, water-cooled oscillating mould. From the bottom of the mould emerges a continuous length of casting to be cut into the length required for further processing. Sir Henry Bessemer patented a form of the process in 1857, his idea being to pour molten steel between two rotating rolls to produce strip or plate, but the concept was never followed up.

It was not until just after the First World War that a form of the process as used for steel appeared. German metallurgist Siegfried Junghans used an open-ended mould for casting brass and copper. Quickly running into difficulties with the solidifying

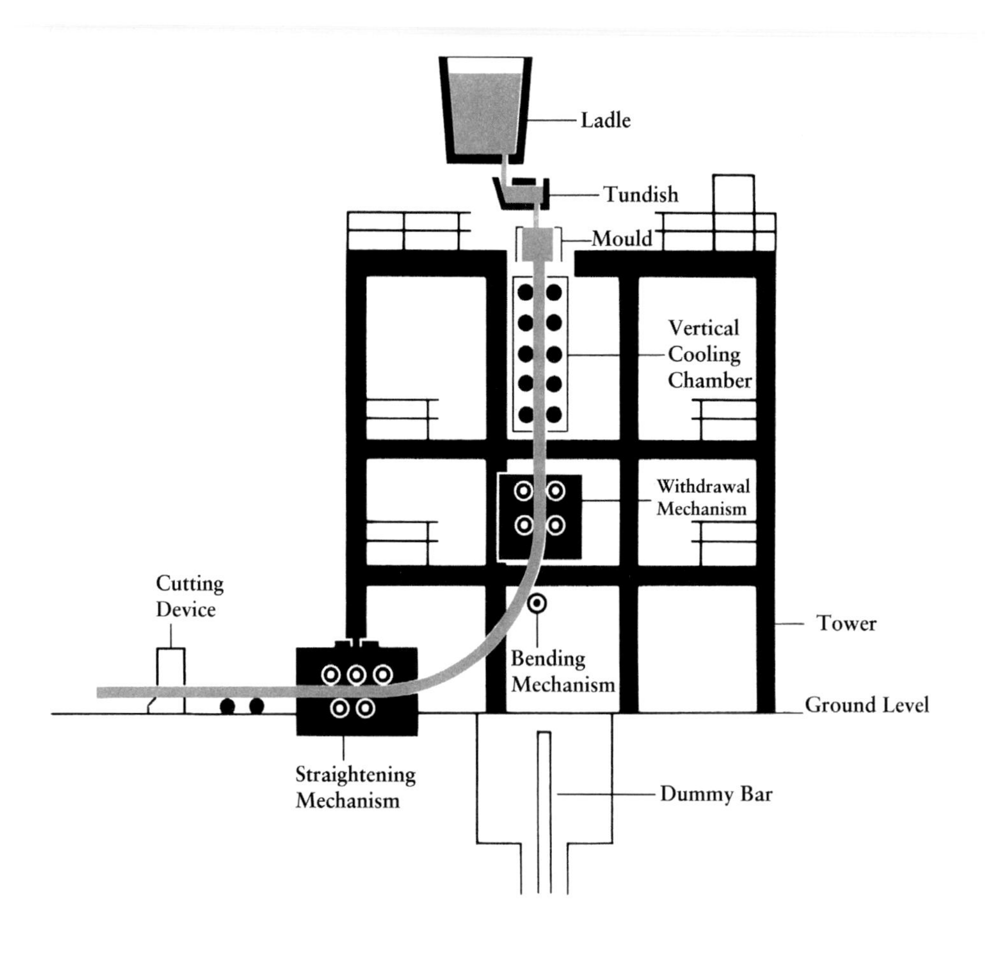

Simplified diagram of continuous casting. (Courtesy of K. Gale)

metal sticking to the sides of the mould, it was arranged to oscillate or reciprocate the mould, which is done today but with a significantly different mechanical motion. In 1933 Junghans patented his process. A few years later while working in Europe as an investment banker, entrepreneur Irving Rossi met with Junghans and immediately saw the potential if the process could be used for steel. He acquired exclusive rights to the Junghans patent in the USA and England. In return Rossi agreed to finance the commercialisation of the process outside Germany.

Meanwhile in Germany, Junghans started experimental work with steel on a 1½-ton plant. During 1949 Rossi was instrumental in setting up a pilot operation in the USA. US steel was, initially, slow to embrace continuous casting and remained happy to stay with the orthodox ingot process. Around this time also a 6cwt plant was started in Austria. Nothing was known at the time of any developments in France or Russia. This, then, as far as was known was the position with continuous casting in 1952 when the Barrow pilot plant became operational.

Process development via works trials was envisaged from the outset. Frank Pearson, a Vickers time served engineer who joined the company in 1940, provided the engineering support. Many problems were involved apart from the sticking issue – for example, the development of techniques for liquid-metal handling, for control of casting and product quality and even the simple question of how to pour liquid steel cleanly into a 2in square aperture mould. By far the most important point was that, for any method to be economically viable, it would be necessary to achieve casting speeds up to ten times faster than any speeds then known. During 1951 contracts were agreed with Rossi for the rights as licensees to use the Rossi-Junghans process.

At the first meeting of the technical steering committee, where terms of reference were agreed, Iain Halliday stated that the process they hoped to develop had been used commercially for some time in the non-ferrous field but progress with steel was negligible. Developments in Britain had not gone beyond the laboratory stage, or emerging therefrom. The 87-year-old works was once again about to be at the cutting edge of steelmaking technology.

The pilot plant was installed at the north end of the old No. 2 steel foundry and consisted of a 5-ton (later increased to 8-ton) basic arc furnace. An electric furnace was the obvious choice, as steelmaking for continuous casting requires tighter control than for orthodox ingot practice where bad furnace operation could be corrected in the ingot mould. In May 1954 Eric Grayson was appointed plant manager and Siemens leading hand Bill Pearson, who had started at the works as a boy in 1912 and was the most experienced furnaceman, joined the squad. Site conditions were so arranged that metal from one of the 40-ton open-hearth furnaces could be brought to the casting machine. It was only after ladle preheating and hot metal handling techniques were developed that trials with open-hearth steel could be undertaken. With the furnace type agreed, ladle designs were the next consideration. Up to this time bottom-pour or stoppered ladles had always been used, but these presented difficulties with flow control and did not lend themselves to preheating as the stopper-rod became overheated. The risk of a 'running stopper' could not be taken. Frank Pearson's technical squad came up with the 'teapot' ladle. This design could be preheated and allowed pouring to continue for up to 1½ hours.

This design of lip-pour ladle was fitted with a 15in deep slag bridge built transversely in front of the pouring lip. The front wall of the ladle sloped up to the pouring lip and the aperture between it and the bottom edge of the slag bridge was about 1¾in by 6in wide. This design effectively held back the slag during pouring – apart from the initial flush of metal – which was disposed of. The ladle lining consisted of two brick courses arranged so that joints were staggered in the radial direction.

It is of interest to note that the preheating arrangements for the ladles was via special burners, the design of which were 'borrowed' from those of combustion cans for jet air-craft. The casting machine tower and associated structures were alongside the arc furnace.

On Tuesday 2 December 1952 the first trial was undertaken with a casting size of 3in square; it didn't work. On 23 February 1953 the first attempt was made at casting a 2in billet, but again it was aborted. These early trials were done with a metal charge of only 2-tons and although not entirely satisfactory, advances were made and experience gained in other areas of the process such as preheating practice and hotmetal handling. Starting to cast with the steel temperature too high resulted in 'breakouts' – with the

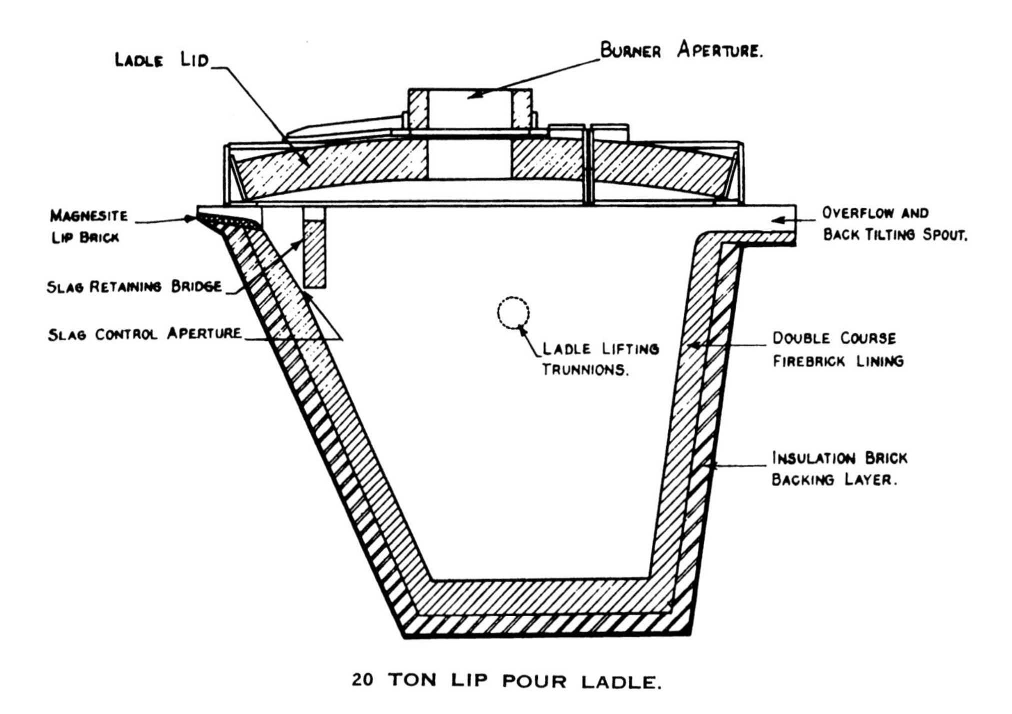

Cross-section of Barrow-designed semi-teapot ladle.

Retirement of arc-furnace leading hand (pilot plant) Bill Pearson, 5 October 1963. Left to right: David Hughes, Roy Millard, Eric Wearing, Bill Pearson, Jack Deakin, Eric Grayson (pipe), Jim Smith, Jim Cheltan, Bill Jones, Danny Smith and Jack Truman.

The mould deck of pilot plant twin-strand casting machine. Top-right is the control pendant while cooling-water pipes to the moulds can be seen left and right. (Courtesy of *North-Western Evening Mail*)

temperature too low the metal would 'freeze' in the ladle before the trial could be completed. The optimum casting temperatures were arrived at empirically and then a series of encouraging trials followed. The desired casting speeds, however, still had not been achieved. It was not until 8 January 1954 (trial 181) that negative strip, which thereafter became known as the Barrow Principle, was first established (negative strip is discussed in more detail in the appendices).

Once the elusive negative strip had been established, faster casting speeds were achieved – 150, 175 and 200in per minute. Then for some reason, breakouts began occurring just below the mould at these higher speeds. Investigations traced the problem to the construction of the mould, details of which had been provided by Rossi.

The Junghans mould was of pure copper 32in long, with a 7⁄16in wall thickness, assembled in two halves giving a longitudinal seam. Although almost a perfect fit, the longitudinal seam, under certain conditions, gave rise to very fine fins being formed on the two opposing faces of the billet. These fins were causing sticking and breakouts just below the mould. The mould was redesigned to be of one piece of solid copper ¼in thick with the addition of a chrome-plated finish to further reduce drag.

Trials with the new Barrow design of mould allowed for greater casting speeds and for the first time it was possible to completely drain the ladle through the machine.

On 31 March 1954 over 1,000ft of 2in billet was produced in one run. On 7 July 1954 a speed of 405 in per minute was reached. This was to be the fastest average speed for casting a 2in billet. Now that production runs could be made routinely with 2in billets, other sizes were trialled – 3in, 4in and 6in by 2in slabs, all proved successful.

Further development work was undertaken in areas such as remote controls, use of aluminium feeder wire into the moulds and the use of mould lubricants such as rapeseed oil. On 6 May 1958 an additional mould was added to the machine to make it a twin-strand unit. By the end of 1958, after a total of 869 trials, it was concluded that the development work had proved successful; finance could now be released for the laying down of a new production plant at Barrow. In the meantime the pilot plant was adapted for development work on large 36in by 5½in slabs in a variety of steels including 18/8 chrome-nickel stainless steel.

During the 1960s Halliday had successes with the continuous casting of various high-alloy steels. This necessitated the shrouding of the metal stream with propane gas for which several patents were applied.

Barrow works now became a training ground for other members of the United Steel group of companies, starting with five steelworkers from the Appleby-Frodingham works at Scunthorpe. This massive plant at the time was the largest integrated steelworks in Europe.

(With acknowledgements to, and paraphrased from, Continuous Casting at Barrow, *by Iain M.D. Halliday)*

The first group of Scunthorpe steelworkers, wearing their protective suits and obviously pleased with their progress, at the Barrow plant in 1960.

This view shows the metal stream leaving the ladle (top) and flowing into the launder where it splits right and left into the tundishes. At bottom left the metal stream can be seen entering the 2in square aperture of the mould, watched by the mould operator.

The second group from Scunthorpe, 13 December 1961.

Taking the bath temperature of the 5-ton arc furnace.

Continuous-cast 9in steel bloom.

Continuous-cast 36in by 5½in steel slab *c.* 1960. At the time this was the largest size produced by continuous casting in the UK.

The mould used for large slab casting. Note the thermocouples for temperature monitoring and the rapeseed oil pipework.

The five original casting machine operatives. Left to right: Isaac Benn, Bill Marklew, David Haney, Bill Caldwell, unknown, Roy Millard, unknown. These men were commended by Iain Halliday for their tenacity in ensuring the success of the process, in his report to the Iron and Steel Institute. (Courtesy of Caldwell family)

9

Barcon

Following an official announcement on the 8 June 1959 the civil engineering contractor Wimpey began preparatory work on the building that would house the new production plant. This building – a smaller version of BAE's Devonshire Dock Hall – would lie on a north-south axis and stand on the site of No. 3 rail mill cooling bank. To provide power to the new installation, Norweb arranged for a spur from the national grid to a substation at Piggy Lane, Hindpool and thence underground to the 10,000kVA transformer house in the works. Wimpey undertook the electrical work associated with the building, English Electric catered for the steelmaking electrical requirements, and Barrow Steelworks and Distington Engineering provided engineering services. The new plant – project Barcon – was designed to produce 36–40,000 tons of steel billets per year. In subsequent years the output was actually 50,000 tons. When plans for Barcon were first issued they were annotated 'Stage 1'; Stage 2 catered for an extension and included the option of resiting the hoop mills into the main steelworks, creating what has become known in the USA as a 'mini-mill'. This, for some reason, never materialised.

Barcon manager Tom McGuire joined the company in 1925 as a youth and worked his way up to become a manager in the Siemens open-hearth melting shops.

As reported in *New Scientist* of 12 May 1960 many countries, including Belgium, France and Russia were keeping a careful eye on developments in Barrow. The report of the 1958 meeting of the Iron and Steel Institute regarding the successful development work and the establishment of negative strip – which was soon to become normal

Heavy going hampers the arrival on site of the first load of steel girders.

operating practice – had aroused a great deal of interest within the industry worldwide. Barrow had made history in more ways than one; as well as pioneering the process it was also the first steelworks to operate without primary mills.

With the initial groundworks complete the first load of structural steelwork arrived on 22 September 1960. The overall construction was completed in a little over twelve months – in fact the plant went into production while some of the services were still being completed.

Barrow was quite late in adopting electric steelmaking (1952) after the first batch of arc furnaces in this country were installed in the Sheffield area shortly after the First World War. The arc furnace had several advantages: it could be used for low-grade, lightweight scrap such as borings and turnings; it could be used to make steel as well as just melt it; and, more importantly, because no fuel was involved, there was no contamination of the metal, which allowed the composition of the steel to be controlled accurately.

The electric arc furnace gave Barrow the most versatile steelmaking tool it had ever had, and then the addition of the oxygen lance gave a further added improvement, especially with regards to time. Blowing oxygen into the bath made the furnace akin to a 'souped-up' Bessemer converter, in terms of removing impurities. Care had to be exercised as overlong use of the lance raised the metal temperature too high, resulting in excessive wear of the furnace lining. The average life of the lining was 193 heats against 281 before lancing.

The old and the new. Erection of the first section with the old No. 2 shed and soaking pits chimney in the background.

The steel framework halfway to completion, 6 December 1960.

Looking south on 5 January 1961. Part of the waste-gas pipeline to the Stirling boilers can be seen in the foreground.

The final cross-member being lifted into place on 20 January 1961. Throughout the entire construction, health and safety had clearly been the main consideration!

Work progressing on the continuous-casting machine with installation of electric motors on the incline roller path in the billet discharge bay, 1961.

The electric arc furnace nearing completion, 29 August 1961.

A full range of qualities was produced at Barrow ranging from dead-soft temper to higher-carbon free-cutting steels. To ensure uniformity of steelmaking, detailed instructions were produced and displayed in the furnaceman's control cabin.

The next photograph shows the oxygen lance/oil burner arrangement mounted on the boiler-plated side of the furnace. Prior to this the lance was portable, being deployed through the working door.

Seat for oxygen lance/oil burner attached to the furnace side.

Refining with oxygen – deploying the oxygen lance via the furnace working door.

While the furnace was producing a heat, the components and equipment involved in the process, i.e the ladle, launder and tundishes were being preheated by special burners to a predetermined temperature. At the furnace, for a given metal analysis, a metal tapping temperature was obtained and checked by immersion thermocouple. When ready the furnace was tapped into the ladle, which was suspended in the tapping pit by the 50-ton crane. Next the charged ladle was transferred to the scales to ascertain the weight of the molten contents.

From here the ladle was lifted onto the top deck of the machine and set into a cradle where safety pins were fitted. The metal temperature was again checked: if too high, a cooling period was allowed; if optimum the combustion-can burner was attached to the ladle lid and ignited. This burner maintained the steel at the required temperature for casting. At this point the mould operators, who were in contact with the ladleman via intercom, turned on the mould cooling water and hung an index wire 14in long into the moulds. Below the mould deck operators were busy feeding the dummy bar up through the withdrawal roll to the bottom of the open-ended mould. This dummy bar was the same cross-section as the section about to be cast. The top of the dummy was fitted with a protruding bolt secured by a tapered pin, which, together with an asbestos washer, closed off the bottom of the mould.

The series of photographs on pp. 102–11 is a pictorial step-by-step excursion through the process.

The mould operators signalled to the ladleman to tilt the ladle forwards via hydraulic controls, the steel established a steady stream into the launder then via the tundish into one of the moulds. (The tundish was so designed to allow one strand to be started initially, and once casting was established the other strand would be started.) The mould operator watched as the metal rose up inside the mould, and when the metal touched the end of the index wire the machine was started. The moulds began to reciprocate and the withdrawal rolls pulled the dummy bar down followed by the newly cast billets into the water-spray cooling chamber. Operators at ground level were warned, via the intercom, to disconnect the dummy bar by knocking out the tapered pins. They next activated the hydraulic bending rolls to direct the new billets through the straightening rolls and along the discharge roller path where they were cut to the required length by flame cutting equipment. The cut billets were conveyed up the roller path until, at the top, they were transferred to a cooling and inspection bank by a tilting-table mechanism controlled by an operative on a control panel.

While casting was ongoing the furnace was recharged for the next heat and the process started all over again. At this point if we pause to consider the word 'continuous', which means without end or unceasing, we see that as far as the process at Barrow was concerned the word was a misnomer. Casting took around two and a half hours, but the furnace took about four hours to produce a melt. Had Stage 2 of Barcon been implemented then it would have been a truly continuous process. On later, larger installations at other works the ladle deck was a kind of carousel called a turret where ladles of steel could be in a queue waiting to cast.

Slagging the furnace prior to tapping. To the left of photograph shift manager Bill Whiteside looks on, 19 October 1961.

Leading hand Tom Lightfoot taking the bath temperature prior to tapping. (Courtesy of *North-Western Evening Mail*)

Final additions prior to tapping.

Taking a metal sample for analysis, overseen by chief metallurgist Mr Joe Lyon. On the extreme left is 50-ton crane driver, Billy Hunter. On the extreme right is furnaceman Ronnie Whitehead.

Pre-heating a ladle in one of the purpose-built pre-heating stations.

Tapping the first heat. Iain Halliday (in dark suit next to safety rail) looks on as the first continuous-casting production plant in the UK goes into operation, 1961. (Courtesy of *North-Western Evening Mail*)

Weighing the charged 'teapot' ladle. The ladle would be weighed again when empty to ascertain the amount of steel processed through the machine.

Lifting the charged ladle onto the casting machine.

Instrument panels for Nos 1 and 2 strands, supplied and installed by English Electric – metals division.

Ladleman taking the metal temperature before starting to cast. It was imperative that the steel was at the correct temperature at the start of casting, 19 October 1961.

A tundish ready for continuous measurement of steel temperature during casting, 26 July 1962.

The ladle, now secured in its cradle with burner attached to the lid, begins to cast.

Mould operators controlling the casting speed.

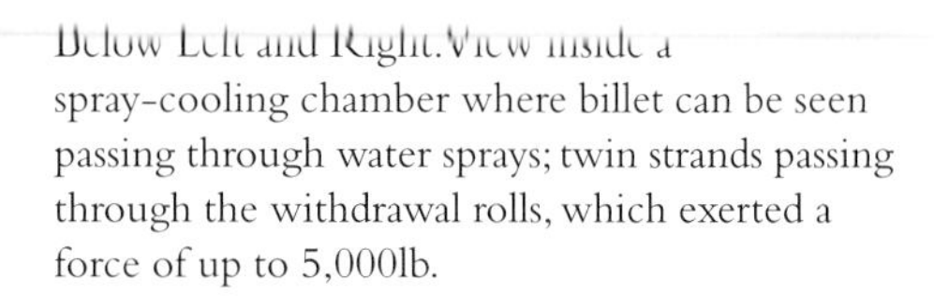

Below Left and Right. View inside a spray-cooling chamber where billet can be seen passing through water sprays; twin strands passing through the withdrawal rolls, which exerted a force of up to 5,000lb.

Billets coming off the machine pass through the straightening rolls, continuing up the discharge roller path.

Cutting the newly cast billet to length.

Billet platform/tilting table with cooling bank beyond.

The cooling and inspection bank with batches of newly cast billets on skids, looking north.

When Things Go Wrong

Once casting was established the process ran almost like clockwork. This was true but on occasion things did go wrong, especially in the early days. The biggest problem encountered was a 'breakout'.

Continuous casting requires that a lot of heat is put in right up to the start of the process, and then as the new steel passes through the water-cooled mould as much heat as possible is removed to allow solidification. During its passage through the mould the billet developed a solidified skin but still had a liquid core. The distance between the bottom of the mould and the withdrawal rolls was the high-risk area where breakouts could occur.

In any event the occurrence could have serious consequences if not contained. Damage to the casting machine and associated electrical and mechanical services resulting in costly downtime, not to mention the risk to operatives. At Barrow rigid process procedures were developed and followed to minimise the risk of breakouts occurring.

Normal operation. Jack Johnson and George Warriner *in situ*.

The same view following a breakout.

A view of the casting machine showing damage to services following a breakout.

Example of 'cobbling' where the descending red-hot billet has failed to enter the straightening rolls and is 'snaking' about watched by a tense Bill Whiteside, shift manager.

10

Fuel and Instrument Department

The Fuel and Instrument Department was formed in 1946 as part of the post-war modernisations and was located initially on the first floor above the Siemens laboratory, adjacent to the Morgan gas producers. About ten years later it was resited in the old electricians' workshop near the Lancashire boilers. The department was formed so that instruments could be fitted to the furnaces and soaking pits to accurately record fuel flows, temperatures and pressures – instead of just relying on the skilled eye of the operators.

Departmental staff in their original building, 24 May 1954. Back row, from left: W.R. Cousins, G. Jones, A. Payne, K.E. Royall. Front row, from left: J. Shepherd, B. Jackson, C. Hooper.

It was deemed essential that daily records were kept so that consumptions could be monitored and wastage kept to a minimum. The department was also responsible for the design, construction, fitting and maintenance of all the thermocouples required for measuring the temperatures in the various areas of the steelmaking processes. The accuracy of all the instruments was also checked daily and if found necessary, repair or recalibration was implemented in the departmental workshop. Boiler feed water was also checked at set intervals and, if necessary, the chemical feed was adjusted to prevent scale build-up on the boiler tubes, which would slow heat transfer.

Departmental staff were designated engineering and not office staff. Although both had similar working hours, finishing time was sometimes dictated by the status of a furnace or the importance of finishing the job in hand. Also, if a serious breakdown occurred outside normal hours of items under the department's control, for instance if fuel trials were in progress (e.g. discovering whether gas oil or propane was the more efficient fuel for preheating ladles), then personnel could be there all night.

An offshoot of the department began in the early 1950s with the introduction of measuring the molten steel when a heat was ready for tapping. This activity employed one person per shift who was solely responsible for the construction and maintenance of the special thermocouples (using platinum and rhodium wire) when required. The departmental manager and most senior member of the team was Ken Royall.

Staff of the Fuel and Instrument Department in 1968. Left to right: Dick Walkden, Ron Crease, Ken Royall and Gordon Ferrier.

Two views of the workshop in the 1970s. The window to the left of the clock looked out onto Walney Road.

11

Sport and Leisure

It will be seen from the number of activities that were available to employees that there was certainly much more to work than work at Barrow. One of the activities set up in the early 1950s, which was open to all employees, was the Camera Club, the most senior member of which was local director, Tom Marple. Meetings were held in the evenings every two to three weeks in the safety and first-aid room, which was spacious and with little obstruction. It also had a projector and screen available, so members could show their own slides or cine-film. Experts from the film manufacturers Kodak

The 'portrait' session in progress. Sitter Diane Fell, cameraman Jack Coward, 3 September 1959.

and Ilford were often invited to give lectures on the different film and photographic papers available, and also to suggest how to improve members' photographs brought in for criticism. The darkroom facility of the Research and Metal Laboratory was also available, so that members could learn to develop and print their own photographs.

After hearing a specialist lecture on how to take indoor photographs using floodlights rather than flash, and the various effects that could be achieved simply by moving the lights to different positions or using different coloured backdrops, a 'portrait circle' was started, which became very popular. Volunteers were sourced to 'sit for the camera' but there was seldom a shortage as they received free copies of all photographs taken. On one memorable occasion the 'model' was Charlie Clayton, a well-known Barrow resident who had embraced the Native American culture for most of his life and had lived with the 'Indians' for quite a long while becoming a full blood brother during his stay.

The Camera Club finally closed its shutters in 1968 when the Instrument and Fuel department had to evacuate its premises (due to many violent explosions from the Barcon waste-gas plant, which inconveniently was sited just outside to the rear).

'Native American' Charlie Clayton in the process of scalping fuel technician Dick Walkden.

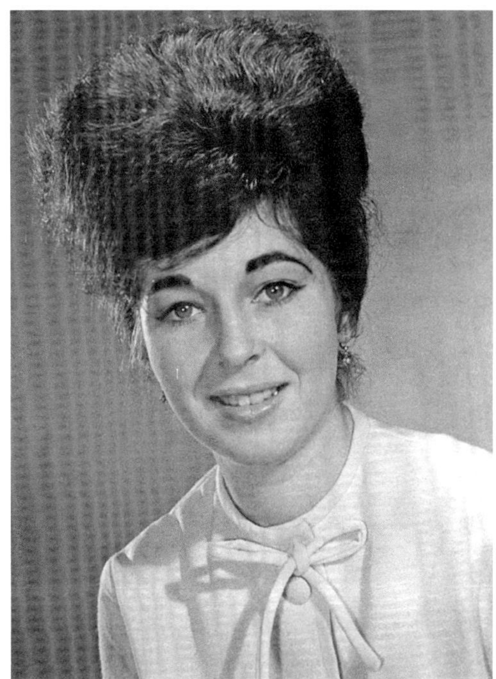

Two subjects of the portrait circle from Barrow works photographed under floodlighting are: Valerie Baker taken 18 March 1959 (Courtesy of V. Baker) and Dorothy Booth taken 6 November 1963.

On the sports front, workers who were so minded had cricket, football, badminton and tennis from which to choose. Soccer was the most popular and as well as departmental teams both the steelworks and Hoop works fielded sides to play in the Thursday League. The Thursday League was set up in the 1920s by and for those whose working week included Saturdays, and so as shops and local small businesses closed early, matches were played on that day at either Croft or Cavendish parks. The police, shop assistants, Pulp works and the GPO all participated in playing for the Kropman Cup. Additionally at Barrow works, interdepartmental knockouts were held.

Tennis matches were played on the hard courts behind the George Hotel on Walney Island, while a hall on the Strand was used for badminton.

Barrow works always had a very good name for its conduct with regards to the employment of children. There were many benefits and subsidies for 'youth' workers on such things as canteen meals, clothing and footwear, and also on time out. One very popular event was the annual summer camp for 15- and 16-year-olds.

The musical interests of workers were adequately catered for by the works brass band that at one time rehearsed in the works canteen.

The band started life in 1884 as the Rising Star Life Boat Crew Temperance Brass Band. The inaugural letter requesting eighteen instruments on loan for the sum of £41 7*s* 6*d* still exists. In 1886 the conductor was Mr Nimrod Wood, an organist at St James church,

Steelworks football team with trophies, for 1956–57, taken on 30 January 1958. Back row, from left: John McKenna, unknown, Bill Chelton, John Griffiths, Ray Broome, Jack Johnson, Ged Ducie, unknown. Front row, from left : unknown, Dougie Williams, Leo Woods, Sid Brough, Ron Speight, Alan Campbell.

No. 2 mill (steelworks) knockout cricket team, 26 August 1956. Back row fifth from left: Ged Ducie. Front row, from left: Norman Barnard, D. Hodgson, unknown, unknown, Bill Wilson, unknown.

Summer camp for 15- and 16-year-olds, 1956. Back row, from left: Jack Charles, unknown, unknown, Jimmy Lynn, Trevor Thompson. Front row, from left: George Martin, George Cavan, Albert Brennan, Charlie Barnard, Frank Holmes.

Steelworks sporting their British Steel logos versus British Cellophane at the Amphitheatre, Furness Abbey, 1973.

Hindpool and an accomplished euphonium player. After struggling to survive, the band was renamed the Iron and Steelworks Band and, with Nimrod Wood still in charge, soon built a reputation, winning trophies in competitions as early as 1890. Perhaps the band's crowning achievement came in 1933 at the Annual Championship Festival, held in Belle Vue, Manchester. Conductor Arthur Baker of Dalton-in-Furness took a very young band to the competition, two members, Jack Bassett, 11, and Stanley Goodall, 12, still being at school. The solo cornet player H. McEnemy was just 19 and the euphonium player H. Bassett was 18. Mr Baker's young bandsmen were up against top brass bands from all over the country and even their firm supporters did not give them much of a chance.

Former Iron and Steelworks Band conductor F.L. Traversi, who was in the audience, said afterwards: 'I have to confess that when I saw the Barrow Steelworks Band mount the stage with all those young boys entrusted to important parts, I felt that, as they were competing against older and more experienced bandsmen, one could hardly expect a rendering that would be equal to the occasion.' They not only held their own against the competition, they played the test piece, Coriolanus, so well that they won first prize, taking the 100 guineas gold championship challenge cup back to Barrow. The fans were delighted, Mr Traversi full of praise. 'The whole performance in every detail was a triumph,' he said.

The Steelworks Band, 1950s. (Courtesy of J. Jefferson)

The Band in Trinity Church Centre, Abbey Road, Barrow, October 2009. (Courtesy of Eileen Wearing)

But the band had troughs as well as peaks during its long history. In the 1960s the band, which had changed its name to the Barrow Steelworks Band when the Ironworks closed, was in the doldrums. Fortunately a few stalwarts kept it ticking over and things gradually improved. Another difficulty was overcome in 1983 when the steelworks shut down. A lifeline was thrown to the band by the manager of the local library, Tony Payne who invited the band round to practice there in the evening, a home enjoyed until fairly recently, when forced out by Cumbria County Council.

The band now practises in the Trinity Church Centre, Barrow with excellent all-round facilities. With a core of dedicated members together with a successful learners' class it looks set to continue for more years to come.

(With acknowledgements to Jim Jefferson.)

12

Subsidiaries

Resume of the subsidiary holdings of the B.H.S. Co. Limited

The mines and quarries division was managed and run from its headquarters at Dowdales Mansion, Dalton-in-Furness.

Coke and Coal

The Barrow Barnsley colliery in Yorkshire raised approximately 500,000 tons of coal annually; the nearby coke ovens produced about 100,000 tons. About 1,700 were employed here.

Iron ore

Park mine near Askam-in-Furness, called the Burlington Pit, was the company's largest source of iron ore. This was the richest and most profitable mine in Furness. Output fluctuated, but about 1,000 tons were mined daily. Around 3,000 hands were employed here. The mine closed in 1921 with the disused buildings being demolished after the Second World War. Stank mine near Roose supplied a smaller quantity of iron ore, but this closed down in 1900. The company co-owned mines at Crowgarth, Bigrig, and also Ullcoats near Egremont, West Cumberland.

The company's annealing ore mill was at Pennington, near Lindal, where ore was crushed and kibbled to required sizes. Thousands of tons were processed here, much being sent out to foundries in the UK.

Limestone

The Crown and Devonshire quarries at Stainton near Dalton were the source of a very pure carboniferous limestone. There was a limekiln on site providing a regular amount of lime, which was an important requirement of the basic steel furnaces. Lime was also supplied to various steelworks in Scotland.

With the subsidiaries and the site at Hindpool, the company employed over 10,000 people, making it, at one time, the largest employer in Cumberland and North Lancashire.

The north pit of the Park iron ore mine and mine buildings at the Park North signal box, *c*. 1890. (Courtesy of Barrow Public Library)

The steel company quarrying limestone at Stainton in the early 1950s. (Courtesy of *North-Western Evening Mail*)

13

Hoop Works Rundown and Closure

The Hoop works had always shown a profit; the high-output 'sweat-box' had customers worldwide. Unfortunately the dated factory of ramshackle buildings, some nearly 100 years old, did not feature in the 1970s rationalisation plans of the British Steel Corporation. In 1962, when all five mills were in production, the combined workforce of Steelworks and hoop works together was 950. The approximate output for that year was 46,500 tons. During 1965, because of the loss of orders in strapping to cold rolling and plastics, and also the changeover in the electrical industry to an increase in the use of solid-core cable and wire armoured cable, Nos 1 and 2 mills were closed down with the loss of fifty jobs. In 1969 – the most productive year of the decade – 48,629 tons of finished products were made and sold, to a value of £1.25 million.

In 1967 just after the nationalisation of the industry the Double-Duo, B-side mill, with a production capability of 6,000 tons per year, was closed down with the loss of another fifty jobs. In its place a new automatic mill was installed. This initiative was against the advice of local management and senior mill operatives who claimed the rolling process would be too slow and therefore uneconomical. After eighteen months of trials and tribulations it was conceded that although rolled product was of good quality the mill could not produce enough tonnage to make it viable. Local management were vindicated, but the mill shut down with more job losses.

Part of the hoop works buildings during demolition in January 2013. (Courtesy of Bill Myers)

The Double-Duo in 1967. Installation of the automated B-side is ongoing, while bottom left a tongsman at A-side finishing rolls is handling matchet steel.

From the start of the 1970s the works still employed 825 people and, despite the fact that No. 3 mill went on short-time working, 40,982 tons of steel was sold in 1970. Up until 1975 No. 3 mill ran at only 60 per cent capacity. In 1975 the overall works staffing levels were down to 650 people and 29,000 tons of steel was produced and sold. No. 3 mill was now working at only 25 per cent capacity, the order book was still diminishing and the outlook was not good. In recognition of this local management asked for a £1 million investment to set up a modernised hoop-rolling facility in the main steelworks near to the Barcon with the intention of producing steel for the orders currently available but at reduced cost (remember Barcon Stage 2?). This was postponed by group management for two years to see if Barrow works could prove viable with the existing plant. British Steel's intention was now becoming clear. The hope that the Brtish Steel Corporation would use its resources to adapt and run Barrow works as a mini-mill was fading quickly.

From a later date, by which time Barrow came under Workington management, No. 3 mill was adapted for rolling matchet steel, and immediately set a new production record by rolling 100 tons in one shift. Steve Thompson became the hoop works' last manager.

Henry Gawron at the shears cutting matchet steel to length in 1967. Henry was one of several Poles who stayed in the country after the Second World War and secured employment at Barrow works. Others included Stephan Frankowski, Janek Dubka and Walter Rutnowski.

Various types of machetes made from Barrow steel. Top centre is a commando knife as used during the Second World War.

Mill manager Alan Johnson presenting flowers to retiring employee Mary Dawson, 23 July 1970.

The re-erected archway at the offices of T. Brady and Son, Transport and Warehousing Ltd. This is located on the old Hoop Works site next to the small reservoir. (Courtesy of T. Brady and Son)

Following demolition of the steelworks' main entrance and office block in 1984, Brady's Haulage and Warehousing Ltd had the foresight and motivation to salvage most of the masonry that comprised the sandstone-arched entrance on Walney Road. They set about reconstructing it on the hoop works site, which they had bought from local entrepreneur Sam Morgan. The sandstone archway became the feature of their offices, which were officially opened on 11 July 1987. The engraving on the commemorative brass plate states: 'This stone arch was originally erected as part of Barrow Steelworks head office which dates back to 1865. The arch was dismantled and re-erected on this site in July 1987 to conserve some of Barrow's steel heritage.'

14

Steelworks Closure and Demolition

It would be easy to assume that the Steelworks rundown ran parallel to that of the hoop works (covered in the previous chapter) but not so. Although downsizing started just prior to the First World War, the actual rundown in terms of plant and machinery began at the start of, or just prior to, the Second World War. Had it not been for the war would the works have closed? The Barrow Haematite Steel Company had insufficient resources to bring the Victorian works into the twentieth century, and entrepreneurship was very much lacking during the 1940s. Way back in the 1890s was the time when investment coupled with a change of direction would have been most beneficial. In 1942 the government intervened and the place became Ministry of Supply, Barrow works. At this point a number of long-overdue but small modernisations were undertaken.

It now had to be faced that Barrow works was geographically disadvantaged; not only had it used up all of the local iron ore but it had no coal in the area either and the isolated location did not now support the existence of a large steelworks. When the development of continuous casting was started in the early 1950s, it was undertaken with the knowledge that a local supply of good quality scrap steel was readily available. This was by way of Thos. Ward, ship-breakers, who operated out of the northern end of Devonshire dock, about 1 mile south of the works. When Ward ceased operations in the late 1960s this posed another problem as more scrap, at a lower grade, had to be brought to the town, thus incurring more transport costs. This being said, the 1960s was a profitable decade for the works, but changes were in the wind.

Following the 1967 nationalisation of the industry, Barrow underwent more organisational changes than it had done during its 103 years of operating. It soon became evident that at Barrow a small operation was ongoing at a very large site. The nationalisation plans of the British Steel Corporation, now renamed British Steel plc, which intended to split up the industry into six product divisions, precluded Barrow works.

Demolition of No. 2 steel shed. The once-prominent bell tower is about to topple, 2 August 1978.

The new scrap bay and handling facility sited west of Barcon, late 1970s.

By the end of the 1970s the three steel sheds had been demolished, and all that stood along the western flank of the site was the Barcon and the corrugated-sheet extension, which housed the billet stocks. The Siemens scrap gantry was decommissioned and replaced with a smaller scrap area serviced by a 10-ton electro-magnetic crane on legs. To cater for the fact that scrap steel was now brought into the works by road, a wide entrance with weighbridge was fashioned to the south of the works entrance on Walney Road.

From October 1968 Barrow joined with the Workington Iron and Steel Company to become the Cumbrian division of the Midland Group of the British Steel Corporation. The Cumbrian division was one of the six divisions into which the Midlands group was divided. Reorganisation of the British Steel Corporation into product groups saw the latest change. On 29 March 1970 Barrow became a part of the new General Steels Division – one of the six new product divisions of the British Steel Corporation. Barrow works was linked with Workington as a unit of the Teesside and Workington Group of the new division. Through an Act of Parliament the various companies who were part of the nationalised industry lost their individual identities as limited companies when the new division came into operation.

The 1970s saw another unfortunate milestone in the history of the works as Barrow became just an annexe of Workington, and Bill Jenkinson, from Moss Bay, was appointed works manager. And so it was that Barrow, which could make any type or grade of steel, became subordinate to a works with no steelmaking capability!

The workforce gradually diminished through the 1970s as many saw the writing on the wall and, thinking there would be more of a chance of finding another opening locally before the job market became flooded later, opted for voluntary redundancy, which was now available to employees.

A view of the derelict works shortly after closure. (Courtesy of *North-Western Evening Mail*)

The demolished transformer house next to the defunct arc furnace, 1984.

Following the official announcement in October the works closed on 23 November 1983. This was reported in the *North-Western Evening Mail* the following day when the paper carried a substantial feature on the historic event. Several senior employees were retained beyond the closure date to facilitate the decommissioning of various items of plant and to provide security until the now derelict site could be made fully secure. Even following closure, orders for matchet steel were being received.

As well as domestic customers, Barrow works' products were exported to all corners of the world. The total value of exports during 1969 was over £1.25 million.

Over the years Barrow steel was supplied to Aden, Australia, British West Indies, Brunei, Burma, Canada, Ceylon, Chile, Egypt, Eire, Falkland Islands, Fiji, India, Iraq, Jamaica, Japan, Jordan, New Zealand, Nicaragua, Pakistan, Peru, Portuguese East Africa, Rhodesia, Singapore, Somali, South Africa, Spain, Switzerland, Tasmania, Trinidad, Tobago, the USA and Zambia.

During the 1970s steel plant capacities began taking on proportions reminiscent of that of the super tankers – once Barrow shipyard had shown such things were possible with the building of the *British Admiral* for British Petroleum – for, just as Barrow steelworks was being dismantled, British Steel commissioned the design and building of a new slab continuous caster at its Port Talbot site in South Wales. This new plant, aimed at stemming the amount of foreign steel coming into the country, had a projected capacity of 36,000 tons per week. This figure was the original design *annual* capacity of Barcon. And that, as they say, is progress!

Epilogue

It is hoped that the foregoing will have given the reader an insight into the scope and size of the operation undertaken together with the achievements at Barrow works. During the years spent researching this book, which involved interviewing and chatting to many people, one common theme emerged – it was a happy place in which to work. Many expressed dismay over the fact that they were involved in developing a process that they taught to others, and which revolutionised steelmaking, and then in just over twenty years saw their jobs go as the works closed in 1983.

Barrow girded the world in steel rails. When billionaire steel magnate Andrew Carnegie came to the town again in 1903, in addressing the assembled dignitaries, he said:

> I came to Barrow as the centre from which I could learn the latest developments in the manufacture of steel. We in America have paid Barrow the most flattering tribute of imitation. We immediately saw what you were doing and we adopted your methods and laboured as hard as we could, our mentor and our model being the works at Barrow. We are your child ...

Notwithstanding, in 1984 an attempt to preserve the works entrance and office block, and also a Victorian water tower, failed despite representations by Barrow Civic Society and a petition raised by locals. Either edifice could have been designed into the new Hindpool Business Park. The question of the office block was referred to a Lancaster firm to evaluate the feasibility and costs associated with preservation, and the Victorian water tower made the front page three times in the *North-Western Evening Mail.*

On 4 October 1984 it was reported that the fight was on to save one of the town's best-known industrial landmarks. It was again reported on 5 October that the 'relic' tower gets a reprieve and then, finally, on 31 October it was reported that in just four seconds the tower came crashing down and was reduced to a pile of rubble.

Despite contributing much to the town over the years, such as land, schools, churches and Barrow's cenotaph, very little remains except disfigured countryside of the great industrial empire the likes of Henry Schneider once controlled, and certainly nothing on the original site save a few misplaced thoroughfares to remind us of a once proud industry.

The Victorian water tower, with demolition of the turbo house well under way.

North-western
Evening Mail
FINAL
16p
26,091 THURSDAY, OCTOBER 4, 1984

Fight is on to save landmark

LAST-DITCH moves are being made to try to save one of Barrow's best-known industrial landmarks, due to be demolished next week.

By Jenny Donnelly

The Victorian water tower (pictured right) on the Iron and Steelworks site, Walney Road, will be knocked down on Monday, unless the decision is delayed, as protesters hope.

The decision to demolish the tower, built in 1864-5, was made yesterday after a site visit by council officials.

They looked at the tower to make an estimate of how much essential repairs would cost after

shame it all came down to money in the end.

"It is a marvellous example of a certain type of industrial style developed in Barrow by people like Furness Railways and the Iron and Steelworks. It is part of Barrow's heritage.

Anyone interested in the campaign to save the tower is asked to ring Steven Donnelly on Dalton 63291.

Front page of the *North-Western Evening Mail* on 4 October 1984 – the fight is on ... (Courtesy of *North-Western Evening Mail*)

Relic tower gets reprieve, 5 October 1984. (Courtesy of *North-Western Evening Mail*)

North-western
Evening Mail
FINAL
16p
26,092
FRIDAY, OCTOBER 5, 1984

RELIC TOWER GETS REPRIEVE

THE HISTORIC TOWER

North-western
Evening Ma
26,114
WEDNESDAY, OCTOBER 31, 1984

IN JUST four seconds, Barrow's Victorian water tower was today reduced to a pile of rubble.

Tower crashes down

The tower, one of Barrow's best known industrial landmarks, had dominated the skyline since 1865, but by 7.30 a.m. today it lay in ruins at the former steelworks (above).

It took just five pounds of explosives to blast the tower to the ground, and demolition expert Tony Thompson said it was one of the least complicated jobs he had ever done.

Earlier this month the council granted the tower a last minute reprieve days before it was due to be demolished after local people called for it to be spared. But after further consideration the council ruled that the tower must go.

The cost of repairing the damaged tower would run into tens of thousands of pounds, it was estimated.

Vickers draughtsman Steven Donnelly, one of the people who was keen to see the tower saved, said: "It is a sad day for Barrow, and I'm sorry that we could not keep the tower. But there are other fine buildings in the town that the council should retain."

He said he felt that the Steelworks offices should be maintained, and that the council should stipulate to the developer of the old steelworks site that the offices should not be demolished, but restored.

It's all over. Tower crashes down, 31 October 1984. (Courtesy of *North-Western Evening Mail*)

The handing-over ceremony of the Barrow cenotaph by Mr M. Ritchie, chairman of the Barrow Haematite Steel Company, to the Mayor, Ald. Fairbairn, 11 November 1920.

Glossary

Brief definition of industry terms used at Barrow works.

Arc furnace	An electric steelmaking furnace.
Arisings	Scrap ends of rails, billets etc.
Basic slag	Slag from basic steelmaking. It contains phosphorus, used as an agricultural fertiliser.
Batch process	The way steel was produced before continuous casting.
Besom	Brush wood, twigs.
Bessemer converter	A pear-shaped vessel in which iron is made into steel.
Billet	A rolled (or continuous cast) piece of steel up to 5in square.
Blast furnace	A furnace in which the primary reduction of iron ore to iron is carried out.
Blast main	The pipe conveying the blast of compressed air from blowing engine to blast furnace or converter.
Blow	The blowing of compressed air into a Bessemer converter.
Blower	The person responsible for operating the Bessemer converter and controlling the blow.
Bloom	A rolled (or continuous-cast) piece of steel above 5in square.
Bosh	The widest part of a blast furnace above the well, also a water tank in a rolling mill in which tongs are cooled.
Bolting-down	A re-rolling term, the equivalent of cogging-down in a primary mill.
Bolting-down rolls	The first pass of a re-rolling mill, the equivalent to a cogging mill.
By-turn	A stand-in or assistant, e.g. By-turn roller.
Campaign	The period of operation of a blast furnace, between blow-in and blow-out.
Carbon steel	An alloy of iron and carbon.
Cobbling	When control is lost of the piece being rolled or continuous cast it is said to 'cobble'.

Cogging mill	A primary mill, also known as a blooming mill. The first set of rolls encountered by an ingot upon leaving the soaking pit.
Cooling bank	An area set aside for rolled products to cool.
Continuous casting	A process for producing semi-finished steel (billets and slabs) from molten steel without primary rolling.
Cupola	A type of small blast furnace used for melting spiegel and pig iron.
Dolomite	A type of limestone used in basic steelmaking.
Dummy bar	A device used in continuous casting to facilitate the start of casting.
Entry side	Front side of a rolling mill where the piece passes into the mill.
Exit side	The side of a mill where the piece issues after rolling.
Fettle	To apply the fettling to a furnace heat, to put right or sort.
Fettlings	The additions added to a furnace lining.
Fire	A local term for a furnace.
Ganister	A powdered siliceous stone applied to a furnace or ladle to give it an acid lining.
Heat	One complete working cycle of a furnace or converter.
Hearth	The lower part of a furnace containing the molten metal (bath).
Hoop	Narrow strip specially rolled for coopers' barrels etc.
Hot saw	Power-driven circular saw located after the exit side of the finishing mill for cutting the piece to length.
Housing	The frame or stand carrying the rolls in a mill.
Ingot	A solidified rectangular steel casting, usually tapered, ready for primary rolling.
Iron	An element (Fe), usually in the form of cast iron, wrought iron or steel.
Joist	A rolled section, formerly called an H-beam.
Ladle	Refractory-lined vessel for receiving molten iron or steel from a converter or furnace.
Ladleman	Worker responsible, in continuous casting, for receiving the ladle of steel onto the casting machine.
Looping	A re-rolling term, where the piece is going through two passes simultaneously.
Launder	Discharge spout of a furnace; also refractory-lined intermediate vessel between ladle and tundish in continuous casting.
Manipulator	A system of rods and linkages used in a mill for manoeuvring the ingots.
Matchet steel	A semi-finished bar sent away to be made into machetes.

Mill furnace	A re-heating furnace for ingots and billets that are to be hot rolled.
Mixer	A large-capacity storage vessel into which molten iron is stored and stirred prior to converting to steel.
Mushroom	Steel bollard set on the floor around which hoop or strip passes during the rolling process.
Pass	Any pair of grooves machined into the barrel of the rolls of a mill.
Pig bed	A flat bed of sand in front of a blast furnace into which the pigs were cast.
Pitman	An operative on the casting pit, usually working with the teemer.
Plate	Wide piece of flat-rolled steel of ⅛in thickness or upwards.
Pyrometer	An instrument for measuring very high temperatures.
Piece	Any piece of steel being rolled was called 'the piece'.
Rabble	A tool used in rabbling (originally used in the puddling process).
Rabbling	To stir or agitate the molten contents of the bath or hearth
Recuperative	A method of recovering waste heat continuously.
Reduction	The process of removing oxygen from iron ore in the blast furnace.
Repeater	A mechanical device used to direct the piece from one mill pass into another.
Re-rolling	Rolling to finished size from slabs or billets (semis).
Reversing mill	A rolling mill that can be run in either direction.
Roller	Senior mill hand responsible for the mill set-up and ensuring that rolled items were to specified finished tolerances. Also, cylindrical 'rolls' set in the mill floor on which heavy ingots etc. could be moved between passes.
Roughing	An intermediate stage of rolling.
Semis	Semi-finished steel, i.e billets and slabs, often passed on to the re-rolling mill.
Sintering	A method of preparing iron ore for the blast furnace.
Skulls	Hollow shells of slag, which are removed from a cooled furnace or ladle prior to relining.
Slab	A rolled (or continuously cast) piece of steel ready for rolling into plate.
Slabbing mill	A primary mill used for rolling ingots into slabs.
Soaking pit	A type of re-heating furnace, at Barrow set into the floor, with lids set on rails similar to a cargo ship's hatch covers,

	and heated by gas or oil. Ingots were stacked vertically and 'soaked' with heat until a uniform temperature was reached.
Stand	A housing complete with its rolls.
Strip	Flat rolled steel up to 18in wide.
Stripping crane	The crane used to separate ingots from their moulds after casting.
Tap	To let the molten contents of a furnace run out.
Teem	To pour molten steel from a ladle into an ingot mould.
Thermocouple	A device with an electrical interface for measuring temperature.
Transfer crane	A crane, also known as a T-crane, used to transfer the ladle of steel from converter to the ingot moulds.
Turn	A man's working shift.
Tuyère	A nozzle at the end of an air blast pipe which projects into a furnace or converter.
Working door	The door on a furnace on the opposite side to the launder, used for working the charge and taking samples and temperature.

Appendix I

Some Key Personnel Over the Years

(in chronological order)

Josiah Timmis Smith	Principal Engineer, First General Manager
Robert Hannay Jnr	Company Secretary
John Rushton	Steel Wireworks Manager
J.M. While	General Manager
A. Butchart	Company Secretary
P. List	General Manager
J. Danks	Ironworks Manager
E. Repton	Steelworks Manager
H. Roberts	Siemens Dept. Manager
A. Timmins	Foundry Manager
J. Clements	Chief Engineer
F. Bowker	Chief Chemist
J. Timms	Rail Bank Manager
T. Prosser	Chief Draughtsman
G. Machin	Traffic Manager
J. Seddon (master builder)	Building Dept. Manager
J. Maltby Black MBE (company stalwart)	Hoop Works Manager
G.N.F. Wingate (colonel)	Steelworks Manager, later Director
J. Lyon MBE	Chief Metallurgist
J.H. Brown	Chief Metallurgist
F.J. Pearson	Works Engineer
Capt. W.F. Baily	Works Engineer

T.G. Marple	Steelworks Manager, later Director
Mr McWhan	Ironworks Commercial Manager
Mr Curnick	Siemens Dept. Manager
J. Killingbeck	Ironworks Manager
R. Stokes	Soaking Pit Manager
E. Ward	Hoop Works Manager
D.W. Ainsbury	Operations Manager, later Director
W. Lockley	Boiler Shop Manager
R.B.Sharp	Ironworks Manager
A. Johnson	Mill Manager (Hoop Works)
E.G. Kite	Commercial Manager
I.M.D. Halliday	Head of Continuous Casting Research
E. Grayson	Continuous Casting Pilot Plant Manager
R. Millard	Pilot Plant Shift Manager
J. Charles	Labour Manager
R. B. Carberry	Assistant Labour Manager
G. Metcalfe	Accounting & Admin. Manager, Co. Sec.
J. Carling	Financial Accounts Manager
K. Law	Production Planning Manager
E.A. Aitkin	Steelworks Manager
T.R. Maguire	Barcon Manager
A. Bull	New Bank Manager
A.E. Wharfe	Chief Electrical Engineer
E. Holme	Works Accountant
C.D.A. Green	Personnel Manager
K.E. Royall	Fuel & Instrument Dept. Manager, (Co. Photographer)
S. Thompson	Hoop Works Manager (last)
W. Roderick	Test House and Independent Inspection
Jim Hollywell	Test House and Independent Inspection
G. McSweeney	Quality Control Manager
Mr Aked	Company Pilot
W. Tippet	Company Chauffeur
D. Cottam	Electrical Manager
W. Jenkinson	Works Manager (last), Workington

Appendix II

Chronology of Significant Events

1854	Hindpool estate bought by the Furness Railway Company.
1858	Hindpool estate leased to Schneider and Hannay.
1859	First blast furnaces blown in.
1864	Barrow Haematite Steel Company formed.
1864	First AGM of the BHS Co. held, 29 March.
1865	Steel company increases its capital from £150,000 to £1 million.
1865	First Bessemer steel made, 23 May.
1866	BHS Co. purchases the Schneider and Hannay ironworks and Park mine for £500,000.
1874	Visit of the Iron and Steel Institute to Barrow.
1893	BHS Co. purchases the steel wire works of Cooke and Swinnerton.
1895, 1896	Temporary closures.
1903	Second visit of the Iron and Steel Institute to Barrow.
1908	Temporary closure.
1917	Visit of King George V and Queen Mary to Barrow.
1920, 1923	Temporary closures, beginning of 'dark ages'.
1942	Ministry of Supply purchases the steelworks and mills from the BHS Co. and run as Ministry of Supply, Barrow works.
1951/52	Nationalisation and subsequent denationalisation of iron and steel industry.
1953	Ownership of the Ministry of Supply works transferred to the Iron and Steel Corporation of Great Britain and subsequently to the Iron and Steel Holding and Realisation Agency. On transfer, the name of the local works was changed to Barrow Steelworks Ltd and the United Steel Companies became managing director. (From 1943 until its closure the ironworks remained under government control and was a separate accounting entity).

1954	Negative strip first established on the Barrow pilot plant, 8 January.
1961	Steelworks officially acquired by the United Steel Companies of Sheffield.
1963	Ironworks bought and closed down by the Millom Haematite Iron Co.
1963	Ironworks closure (31 March)
1983	Steelworks closure (23 November)

Appendix III

Anecdotes, Trivia and Miscellany

Contrary to reports in some areas of the media, the steelworks at Barrow did not shut down during the Second World War. It was quite active in support of the war effort. The mills were kept going by young women who filled in for the men away on active service.

Ken Law, production planning manager, started 1936, retired 1980.

From the standpoint of the man in the street, it would appear that the town's largest heavy industries, with their sprawling size, large black workshops, steam, smoke and noise would operate along very similar lines. Quite the opposite! They both ran with different cultures. There was always something of a 'friendly' rivalry between the two industries. Shipyard workers would say that the steelworkers were paid from the neck down. Steelworkers parried with the remark that the 'yard' carried more passengers than British Rail. One example of the different culture was in overhead crane driving. There was an overhead gantry in No. 2 shed, which extended from the north end to about two-thirds of the way down the building. On it were three cranes used to service No. 2 rail mill. The northernmost was a 10-ton swl [safe working load] Wellman called the Klondike crane, next was a lighter, faster crane called the Whip. Thirdly was another 10-ton Wellman referred to as the south soaking pit crane. When the mill was on the Whip would be racing up and down the bay, feeding the cogging mill with white-hot ingots from the soaking pits. On the other hand, in the shipyard, either in the Gun shop or Assembly shed for example, a crane driver may only get one lift per day. On one occasion a shipyard crane driver came to work at the steelworks. He lasted one shift – they put him on the Whip. His leaving remark was that he didn't come to be a racing driver wearing dark glasses!

Peter Keenan, soaking pit crane driver (50 years' service).

Young woman operating the mechanical manipulators on No. 2 mill *c.* 1955. This was a typical sight during the Second World War. (Note that there was no gender pay gap at Barrow works.)

On the afternoon shift during the 1940s, one of the leading hands on No. 3 mill at the hoop works would go around asking who wanted some beer. Someone would then be sent out across Wilkie Road to the Soccer Bar (which was under the stand of Barrow AFC in those days) for jugs of ale. 'Drink as much as you want' was the unofficial policy – but keep the mill going. This surprised me, given the number of mutilations I had seen. The place was a hellhole in those days.

Jim McGlennon, No. 3 mill furnaceman.

I left the steelworks in 1950 and went to work in the Buccleuch Dock plumbers' shop at the shipyard. It took a while to adjust to the culture there. It was a mortal sin to be caught sitting down in those days. The foreman was very petty. If someone had a brew of tea on the go outside of the official ten-minute break, the foreman had been known to knock it over. At the steelworks you sat down with a cuppa when you wanted – but you never abused it.

Frank Rogan, plumber, Gas and Water dept.

In the early days of the steelworks, Mr Smith had an overseer, a Mr Sutton working for him on the Bessemers. He rode to work on a jet-black horse with a solid silver harness. Mr Sutton later bought the Wheatsheaf Hotel in Hindpool and two other town-centre public houses.

Alice Leach, The Religious of the Sacred Heart of Mary.

It has long been known that the steelworks directors contributed greatly to the construction of Hindpool Parish Church, St James the Great and that, prior to the church being built, Sunday services were held in the work's Pattern shop. Such was the director's concern for their worker's spiritual welfare but … St James church was so named apparently because of the influence of one James Ramsden and his acts of philanthropy that not only extended to providing a place of worship but also attending services to ensure that those who were granted a Sunday off to worship, actually attended. Otherwise it was 'six days shalt thou labour, but on a Sunday ye shall do a double shift'!

John Saddler, Cook Street, Hindpool.

I was almost 19 when the Second World War broke out, and women without children were required to register for war work at the local Dept. of Labour. I was called up in 1941 and directed to Barrow Steelworks, as were many girls of my age. The steelworks was a truly dreadful place to work in! Talk about 'Dante's Inferno!' Heat, dust and bits of discarded metal lying about the red stained floor made it a real hazard to walk about on. We all sat operating the rollers set in the floor. At the far end was a Siemens furnace, and when it opened, a blast of white-hot heat travelled the entire length of the mill. The furnacemen, stripped to the waist, grappled the huge glowing ingot to get it onto the first set of rollers, which were operated by a man in the crane above, which sent it to the first huge press, showering white-hot sparks which danced like fireflies in the gloom. The press squeezed it as effortlessly as though it were a lump of butter, so that it came out the other side narrower but longer and dropped it onto the next set of rollers. Now it was our turn to operate our levers and send the red-hot monster slithering along the rollers to the next pressing to be operated by the two girls in the next section. After it passed, the sweat rolled off us and we had to struggle into our coats to avoid getting a chill. The rolling mill was near Walney Channel, from where chill winds swept through the gaps in the walls of the Victorian building. This whole operation was repeated the entire length of the mill, until it became a steel railway line. The last operator worked the billet shears, which trimmed off the jagged edges and then sent it rattling away to the railway trucks outside.

Katy Percy, b. 1920, Barrow.

An attractive feature of the steelworks in the 1960s and 70s was that small orders, say 17 to 25 tons of a special steel, could be received into the works and within the week the steel could be made, rolled and despatched. This was not possible before continuous casting.

Jim Ducie, ex-steelworker, McClintock Street, Hindpool.

The typing pool, Christmas Eve 1958. From left: Dorothy Stamper, Ethel Lowe, Maureen Dent, Winnie Trotter, Jean McKinnon, Alma Jones (seated), unkown and Barbara Woods.

Undated photo of the slag-bank loco at the start of its run. Note that the driver (Bob Ritson) appears to be wearing a tie!

My very first recollection and maybe the funniest to me, was told by my Dad, Stan Henderson snr, about 'Uncle' George who was one of the drivers of the slag bank steam locomotive. Barrow slag bank was said to be one of the longest in Europe at one time. Around 1959 the company was changing from steam to diesel and was in the process of being supplied with new engines from Hunslet in Yorkshire.

Prior to making the purchase, a representative from Hunslet came to assess the Barrow works' requirements. At his first visit Uncle George invited the man to accompany him on the footplate on a routine slag run. Halfway up the slag bank the Yorkshireman seemed to go white and insisted on disembarking. When asked why by my uncle he firmly stated: 'Nay lad, I never ride an engine where the seagulls fly lower than the train tracks!'

Stan Henderson, Hood Street, Hindpool.

Acid or Basic (ref. Chapter 6)

Iron and steel in the molten state will readily react with certain elements and the outcome may be desirable or not as the case may be. One of the reasons that coke, not coal, is used in a blast furnace is because the molten iron would take up unwanted sulphur from the coal. Coke is coal with the tar and sulphur removed. Furnaces and ladles are insulated from their molten contents to not only protect them from the intense heat but to influence the type of steel made. The insulating material used is a firebrick (refractory) lining. This lining material can be either chemically acid, basic or neutral. Acid steel therefore is made in a furnace that has an acid (siliceous) lining, and also under an acid slag, which the furnaceman controls. Ganister, made from a type of siliceous crushed sandstone, and applied by trowel, gave the Bessemer converter its acid lining. Similarly, in a furnace making basic steel the lining would be 'basic', e.g. Magnesite (dolomite for instance). The furnaceman would shovel crushed dolomite and lime into the 'fire' to make a floating white slag. During steelmaking the industrial chemists were always in evidence taking regular samples to ensure the 'heat' was to specification.

Negative Strip (ref. Chapter 8)

At Barrow the mould was reciprocated vertically as done by Junghans but with a significant difference, Barrow's mould motion became known as the Barrow Principle. Negative strip allowed for sustained casting speeds above 200in per minute. The mould was moved up and down by means of a cam (eccentric) and link mechanism driven by a 5hp electric motor. The reciprocation cycle was such that on the down stroke the mould travelled slightly faster than the newly cast section, and then returned on the upstroke at about three times as fast as the descent speed. (Because the foregoing may not easily be visualised, and not conveyed by still photographs, cine-films of the process in action were shot at Barrow in 1956 and 1958, and these films reside with the BFI). This slightly faster descent speed of the mould was the main characteristic of the Barrow Principle, which eliminated false-wall formation as the newly cast billet solidified against the mould surfaces. Negative strip over a stroke length of 2in was between 1⁄32 and 1⁄8 of an inch. For all casting speeds, the movement of the billet or casting can be superimposed on a diagram of mould movement, which basically is the profile of the cam. This is done on the diagram shown below. The line ABC is the mould movement to a time baseline of one complete reciprocation cycle, as represented on the cam profile. Between A and B the mould descends through its own stroke length at nearly constant rate for three-quarters of the cycle, whilst between B and C the upstroke occurs. The dotted line PP represents the superimposed descent of the billet, if conditions are synchronised, as is done when operating on the Junghans principle. The line RR represents the billet descent when slower than the mould. Since these lines coincide at the beginning of mould descent, the difference at the end is real and, as indicated, is known as 'negative strip'.

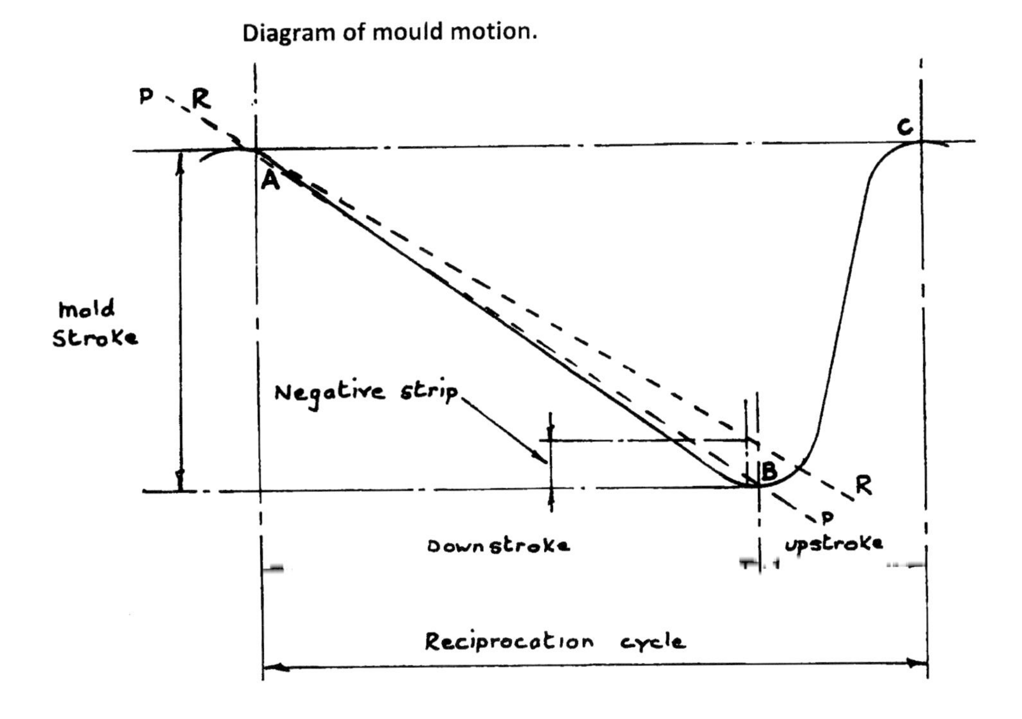

Diagram of mould motion.

Mr E.A. Aitkin (right) marking the retirement of steelworks manager Tom Marple on 30 June 1964 at the Fisherman's Arms, Baycliff.

Time office staff *c.* 1918.

Technical staff gather for the retirement of Frank Huddleston. Works engineer Frank Pearson is making the presentation. Those known are, from left: Jack Smithson, unknown, Alan McDougal, Dorothy Greaves (tracer), far right is Dick Carr, 22 September 1961.

It was quite common for whole families to work in the iron and steelworks. In fact many sons followed in fathers' or older brothers' footsteps. Below is a collection of siblings who did just that:

Len Chelton	electrician	Fred Parry	merchant mill (boss)
Ernie Chelton	crane driver	W. Parry	rail tester
George Chelton	labourer	Bert Parry	charger driver
Ged Ducie	crane driver	Matt Shannon	chisel sharpener
Jim Ducie	slinger	Tommy Shannon	elec. labourer
Jack Henderson	plate mill	Charlie Shannon	rail bank
Johnny Henderson	No. 2 mill	Billy Shannon	No. 2 mill
Stan Henderson	Barcon		
Billy Henderson	Barcon		

Eric Whitehead	rigger	Ronnie Whitehead	furnaceman
John Whitehead	labourer	Edwin Whitehead	bricklayer
Mike Haffner	hoop works	Jack Haffner	hoop works
Gerry Haffner	hoop works	Harold Baines	hoop works, roller
Dickie Baines	hoop works		No. 2 mill

The retirement of Mark Drinkhall (electrician) with his colleagues in the electrical department on 18 March 1968. From left: Mr A. Wharfe, L. Seward, P. McKenna, Ray Millard, W. Parry, W. Wilson, M. Drinkhall, R. Hibbert, P. Gillen, M. Greaves, T. Hooper, D. Edwards, Mr D. Cottam, J. Huitson and Wilf Bethwaite.

Steelworks Office Staff outing to Harrogate. 3-6-1950.
(Personnel from left to right)

Back Row. Ray Carberry, Leslie Parkinson, Alan Bull, Harold Carling, Frank Taylor, Maynard Wade, Alf Whittam, Maxie Dunn, Dick Hines, Alf Roscoe, Marshal Harper, Ken Wearing, Charlie Jackson, Des Delaney & Cyril Taylor.

3rd Row. Harold Smith, Edna Tippins, Audrey Coulton, Tommy Johnson, Barbara Griffiths, Audrey Jones, Jean Paterson, Maud Nash, Joan McWilliams, Alma Jones, Brenda Kent, Pearl Thomson, Sylvia Collins, Janet Poznet, Pat Thompson, Jennie Flemmings, Dorothy Clarkson, Jock Rankin & Don Greenwood.

2nd Row. Hilda Wanlass, Connie Brown, Sister Beattie, Eileen Blundell, Wilf Bradshaw, Phil. Howgate, Ted Kite, Colonel Wingate, Geoff. Metcalfe, Harry Bailey, Jimmy Conlin, Winnie Trotter, Kathleen Holmes & Joan Heffernan.

Front Row. Alan Sandford, Ron Metcalfe, Tommy Hill,Sheila White,Elsie Haslam, Derek Wilkinson, Jack White, Jack Yates, Winnie Wren, Joan Briley, Jean Main & Joe Shields.

Office staff outing to Harrogate, 3 June 1950.

Ramsden Square: Office staff outing to the races on 26 May 1956. Barrow with Lakeland Laundries factory is in the background.

Bibliography

Works consulted in preparing this book

Banks, A.G., *H. W. Schneider of Barrow and Bowness* (Kendal: Titus Wilson, 1984)

Barrow Haematite Steel Company, *Barrow Steel: A Brief History and Survey of Productions* (Burrow, Ed. J., 1937)

Byers, R., *Workington Iron and Steel* (Tempus, 2004).

1901 Census for England

Clarke, James E., *Barrow Works – A Unique History* (British Steel Corp.)

Gale, K., *The British Iron and Steel Industry* (David & Charles Ltd)

Gale, K., *The Continuous Story of Continuous Casting* (British Steel Corp.)

Halliday, I.M.D, *Continuous Casting at Barrow* (Barrow Steelworks Ltd, 1958)

Hewson. R., *Let's All Go Down the Strand* (M.A. and R. Hewson, 2004)

Jackson, A., Lyon, J. & McGuire, T., *Continuous Casting of Small Billets at Barrow Steelworks* (Iron and Steel Federation, 1964)

Leach, Alice, *Our Barrow, Part 3* (Alfred Barrow School, 1981)

Leach, Alice, *Voices from Barrow and Furness* (The History Press, 2008)

Marshall, J.D., *Furness and the Industrial Revolution* (Barrow-in-Furness Library, 1958)

Myers, Bill, *Barrow-in-Furness Remembered* (Tempus, 2000)

Report of the visit of the Institution of Mechanical Engineers to Barrow works, 1880 (Grace's Guide)

Totten, G.E. & Co., *Handbook of Metallurgical Process Design*, 2006 edition (New York: Marcel Dekker Inc., 2006)

USC Ltd, *This is United Steel* (USC Ltd)

Papers and Periodicals

The Engineer, 4 September 1874

Iron and Coal Trades Review, 4 August 1899

Barrow News, 18 January 1963

North-West Evening Mail, 4 January 2014

Websites

news.bbc.co.uk (BBC-h2g2-Henry Bessemer and the Development of Bulk Steelmaking)

Sources by chapter

Chapter 1: Hindpool

Let's All Go Down the Strand, Hewson, R.
Alice Leach, personal communication
John Binnel, personal communication

Chapter 2: The Hindpool Blast Furnaces

Barrow News, 18 January 1963
North-Western Evening Mail, 2 November 1963
Albert Brennan, ex-Barrow Ironworks
George Hurley, ex-Ship Drawing Office, Vickers, Barrow

Chapter 3: The Birth of Bulk Steelmaking

news.bbc.co.uk (BBC-h2g2-Henry Bessemer and the Department of Bulk Steelmaking)
The Engineer, 4 September 1874
Iron and Coal Trades Review, 4 August 1899
Furness and the Industrial Revolution, J.D. Marshall, 1958
Furness Past and Present, Div. 5, J. Richardson
1901 Census for England
W. Pearson, ex-steelworker, 1912–63, personal communication

Chapter 4: Rolling Mills and Soaking Pits

Jack Hool, personal communication

Chapter 5: The Hoop and Bar Mills

Rushton family history
Frank Pearson jnr, personal communication
Billy Miller, BSW 1950–82, personal communication
Cumbria Steel, 4 October 1968

Chapter 6: Open-Hearth Plant and Steel Foundries

H.W. Schneider of Barrow and Bowness, A.G. Banks, 1984
Iron and Coal Trades Review, 4 August 1899
Barrow Steel: A Brief History and Survey of Productions, BHS Co., 1937
W. Pearson, BSW, 1912–63, personal communication

Chapter 7: Boilers, Engines and Transport
Iron and Coal Trades review, 4 August 1899

Chapter 13: Hoop Works Rundown and Closure
North-Western Evening Mail (various dates)
Jim McGlennon, Union Representative – Iron, Steel and Kindred Trade
ISTC Banner, 15 February 1982

Chapter 14: Steelworks Closure and Demolition
Barrow Works, A Unique History, James E. Clark
North-Western Evening Mail (various dates)